Plain

About

Drinking
Water

Fifth Edition

WITHDRAWN

Learning about Water and Energy Conservation

**Generously funded by
Riverside Public Utilities**

WATER | ENERGY | LIFE

CITY OF
RIVERSIDE

PUBLIC UTILITIES

Plain Talk

About

Drinking
Water

Fifth Edition

Answers to Your Questions About the Water You Drink

Dr. James M. Symons, Author Emeritus

Gay Porter De Nileon, Editor

Allen Panek, Jackie Glover, Contributing Editors

**American Water Works
Association**

"Weird, Wonderful Water," reprinted by permission from The Young Naturalist: The Wonder of Water, ©2005 by Mary Hoff, published by the Minnesota Department of Natural Resources.

Lyrics from "We All Live Downstream" used with permission. Copyright © 2008 Banana Slug String Band, "We All Live Downstream"

American Water Works Association
6666 West Quincy Avenue
Denver, CO 80235-3098
303.794.7711

Interior Design and Production: Fran Milner, Hop-To-It Design Works
Cover Design & Illustrations: Melanie Yamamoto

Library of Congress Cataloging-in-Publication Data

Symons, James M.
  Plain talk about drinking water : answers to your questions about the water you drink / James M. Symons ; Gay Porter De Nileon, editor ; Allen Panek, Jacqueline R. Glover, contributing editors. -- 5th ed.
     p. cm.
  Includes index.
  ISBN 978-1-58321-742-9
  1. Drinking water--Miscellanea. 2. Water quality--Miscellanea. I. Porter de Nileon, Gay. II. Title.
  TD353.S96 2010
  363.6'1--dc22

                         2009035953

ISBN 1-58321-742-8
ISBN 978-1-58321-742-9

Table of Contents

Preface

Nine years have passed since the last edition of *Plain Talk about Drinking Water* was published. Since that time, many things have changed in the water industry. Regulations have become stricter. Detection, monitoring and treatment technology have advanced substantially. Bottled water has become increasingly more commonplace. Most importantly, consumers have become better educated and aware of the products—and the water—they use and consume. Consumers today are concerned not only with where their water comes from and what's in it; they want to know more about how to protect it from contamination, how to conserve it and how to treat it in their home. They want to know about bottled water and about pharmaceuticals in tap water. They have concerns about the security of water supplies and question the effects of climate change. They don't want to be placated with simple answers; they want the truth told plainly, in language they understand but that doesn't talk down to them.

With both the novice and the sophisticated consumer in mind, this new, fifth edition of *Plain Talk about Drinking Water* has been thoroughly revised and updated. Beginning with the strong foundation of the first four editions that contained common questions answered by Dr. James M. Symons, we researched, queried and vetted all the material—updating, adding, rewriting and clarifying answers to common, and some not-so-common, questions about the water we drink.

With more than 200 questions and answers, this new edition covers a wide range of topics, all with the common theme of water, that most vital element of life. The first chapter, Health, is the largest, filled with facts about water's contribution to health, common chemical and microbial contaminants, treatment chemicals, travel precautions and other information that relates to the central question, "Is my water safe to drink?" Subsequent chapters provide information about:

- Taste, odor and appearance—aesthetics of tap water
- Home facts—ways water is used in a home as well as information about bottled water, home treatment and costs
- Sources—groundwater, surface water, reservoirs, pollution issues and source water protection
- Distribution—pipes, hydrants, tanks and leaks
- Conservation—the why, what and how of saving water
- Regulations, testing and security—the role of the federal, state and provincial governments, and what's being done to protect water supplies from intentional harm
- Fascinating facts—beyond drinking water, from solid to liquid to gas

The appendices include 1) acronyms used in this book, 2) a list of all the questions in the book so the reader can easily find related information, 3) details about natural chemicals found in source waters, and 4) a pronounciation guide to some of the more unusual terms discussed in this book.

Our goal is to inform, entertain, educate and, along the way, provide you with a greater appreciation for the element that is so essential to our lives—water.

Gay Porter De Nileon
Editor
November 2009

Acknowledgments

Many books, journals, newspapers and Web sites were researched, and numerous experts were consulted in the process of updating this book. The most useful resources were those from official and authoritative sources, including American Water Works Association, U.S. Environmental Protection Agency, U.S. Food and Drug Administration, U.S. Geological Survey, U.S. Department of Health and Human Resources, Health Canada, Environment Canada, Centers for Disease Control and Prevention, National Academies, Water Research Foundation, World Health Organization, Water Quality Association, NSF International, Underwriters Laboratories, Canadian Bottled Water Association, the Nuclear Regulatory Commission, The Mayo Clinic and many peer-reviewed science journals, professional publications and general news articles.

Every effort has been made to make this material as accurate and current as possible; however, we do acknowledge that there may be some information that researchers and others will dispute. So much information exists about water that this book can only cover a minimum about some issues. We encourage our readers to investigate topics of interest more thoroughly on their own and draw their own conclusions.

This book would not have been possible without the assistance of many people the editors would like to acknowledge here: Fred Bloetscher, Gary Burlingame, Jack Hoffbuhr, Greg Kail, Pat Kline, Cynthia Lane,

Bill Lauer, Jeff Oxenford, Alan Roberson and Steve Via for their reviews and comments on various sections of the manuscript.

Thanks also to Nicole Durfort for her contributions toward the updating of Canadian plumbing codes and drinking water regulations; Reid Campbell, Ray Bilevicius, and Duncan Ellison for their expertise on many other Canadian issues; David S. Panek for his insights into the world of home aquarium operations; Marguerite K. Panek for her tireless efforts in preparing draft copies for electronic transmission; Allan L. Poole, P.E., for his timely and accurate redrafting of items about home use; Dr. James Symons for the original manuscript and many useful news articles that prompted new questions and answers; Tom and Maureen Hodgkins for help in getting the latest information about water conservation; Janice Kaspersen of Forester Media for substantial insight into stormwater issues; James E. Holzapfel, P.E., for information about distribution systems, and, finally, to Martha Ripley Gray and Jen Reeder for their eagle-eyed proofreading and to Fran Milner and Melanie Yamamoto for making the book look so good.

Weird, Wonderful Water

Water is as solid as rock, but light enough to float. It's a gas that turns to liquid when it touches cold air. In liquid form, it defies gravity. It's one of the strangest chemical compounds ever discovered. Yet it's the most common substance on Earth. It's found on every continent in the world. It's in the sky, in the oceans, in your backyard, in your house—even in you. It's weird, wonderful water.

Water is almost everywhere, so we're pretty used to it. In fact, you might think of it as a little boring. But it's not. Due to some chemical quirks of nature, water behaves differently than other substances. And that makes all the difference in the world. Plants and animals depend on water's odd traits to carry out the everyday business of living. If water weren't so strange, life as we know it would not exist.

—Mary Hoff, *The Wonder of Water,* 2005

Editor's note—Many topics are covered in several different sections of the book. For example, point-of-use treatment devices are discussed at length in the *In the Home* chapter, but the devices are also discussed as treatment options in the *Taste and Odor* chapter and in the discussion of chemical contaminants in the *Health* chapter. So readers should refer to the index in the back of the book or master list of questions in Appendix B to learn if there is more information about a particular topic in this book.

Health

I believe that water is the only drink for a wise man.
— Henry David Thoreau

1. Is my water safe to drink?

Tap water from public water systems in the United States and Canada is among the safest in the world, and maintaining that quality is a priority for the U.S. Environmental Protection Agency (USEPA), Health Canada and public water utilities.

News stories often discuss one contaminant or another found in the water we drink, and sometimes short-term disease outbreaks are traced to water supplies. However, most water utilities monitor for and control more than 100 different parameters that may affect water at the tap—from algae in the source water to lead in homeowners' pipes. Regulatory agencies in Canada and the United States set drinking water quality standards and guidelines for community water systems that require this level of diligence. You'll find that most water professionals consider themselves stewards of public health and safety who also drink the water that is delivered to homes in their communities.

In the United States, the Safe Drinking Water Act sets the standards for drinking water treatment; in Canada, *Guidelines for Canadian Drinking Water Quality* provides the basis for drinking water quality parameters. In most cases, USEPA standards and Canadian guidelines for a particular contaminant are the same.

2. Is potable water the same as safe water?

When it is safe to drink, water is called *potable*, which rhymes with "floatable." Water is considered safe to drink if it meets or exceeds all of the federal, state, and provincial standards that are legally enforceable. In the United States, if your tap water does not meet any one of the standards, your water supplier must notify all its customers of the problem.

3. How much water should I drink to stay healthy?

A report issued by the Institute of Medicine of the National Academies says that most healthy Americans typically allow their thirst to guide them rather than following the old "eight to nine glasses a day" rule, but thirst alone should not be used as a guide for when to drink. By the time you become thirsty, it's possible you may already be slightly dehydrated. Others would say the amount of your water intake should depend on your weight, where you live and how active you are. Because 60 percent of your body is water, it is essential to replenish what water you lose through your breath, perspiration, urine and bowel movements, which for most people is about 8 cups (2 liters) a day. The Mayo Clinic recommends:

✓ drinking a glass of water with each meal and between each meal,

✓ hydrating before, during and after exercise, and

✓ substituting sparkling water for alcoholic drinks at social gatherings.

Older people should not rely completely on thirst to determine their need for liquid because their thirst mechanism may have lost some of its ability. Consumption of salty foods, diseases such as diabetes and various medications can all affect a person's thirst sensation. Finally, in proportion to body weight, babies need more fluids than adults. Consult with your doctor about the water needs of your baby.

4. Must all my water intake be plain water or do drinks made with water count?

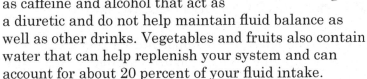

Juice, milk and soft drinks are almost all water, so they do count toward the required total daily fluid intake. Nutritionists often recommend tap water, however, because some other beverages contain chemicals such as caffeine and alcohol that act as a diuretic and do not help maintain fluid balance as well as other drinks. Vegetables and fruits also contain water that can help replenish your system and can account for about 20 percent of your fluid intake.

5. Does drinking water contain calories, fat, sugar, caffeine or cholesterol?

No. Water is the original "health" drink.

6. Can drinking water help me lose weight?

Most weight loss programs call for drinking healthy amounts of water every day to help the body process food and counteract hunger attacks. Often we feel hungry when we are dehydrated; by drinking a glass of water, we will be satisfied without consuming any calories!

7. Does drinking ice water help burn fat?

While your body does have to work a little bit harder to raise the temperature of the ice water to body temperature, it would take more than a year for a person to lose one pound of fat by drinking a glass of ice water every day.

Do the math: An 8-ounce (0.24-liter) glass contains about 240 grams of water. Water with ice is at a temperature of 32 °F (0 °C). When consumed, the water warms to body temperature in the stomach, about 98.6 °F (37 °C). It requires about 8,800 gram-calories (240 × 37) to warm the water. Food calories are actually kilogram-calories, so drinking the glass of ice water uses 8.8 kilogram-calories. Thus, very little fat is burned in food energy terms.

8. Is there a connection between drinking water and healthy skin?

Drinking enough water is important for the health of your skin. Staying properly hydrated removes the toxins that build up in your skin and helps prevent it from drying out.

9. How long can a human go without water?

This depends upon weather, shelter, altitude, activity level and whether a person has access to food that contains water. For example, in hot conditions with no water, dehydration can set in within an hour and death can occur within two to three days. An infant left in a hot car or an adult exercising hard in hot weather can dehydrate, overheat and die in just a few hours. In some conditions, a person can live up to a week to 10 days without water, but water deprivation isn't something that should be tested! On the other hand, most adults can live three to six weeks without food, depending on how much fat is stored in their bodies at the onset of food deprivation.

10. Is it true that you can drink too much water?

Yes. Drinking too much water is uncommon, but water intoxication can be serious. Too much water intake can result in hyponatremia, which occurs when the sodium in your blood is diluted. This imbalance of electrolyte content makes the body's tissue cells swell, which puts pressure on the brain and nerves and can cause an irregular heartbeat, fluid in the lungs and fluttering eyelids. Seizures, coma and ultimately death can occur unless water intake is restricted and a hypertonic saline (salt) solution is administered.

Endurance athletes, babies and some drug users are potentially at risk from this condition. Athletes who sweat heavily may suffer water intoxication if they drink water without added electrolyes. Babies are susceptible if they drink too many bottles a day of diluted infant formula. Finally, the illegal drug Ecstasy (MDMA) can cause the body to overheat and subsequently induce water intoxication to compensate. Numerous Ecstasy users have died of hyponatremia.

11. Does anyone actually get sick from drinking tap water?

Yes. If a water system is compromised in some way, it can become contaminanted. Drinking water containing viruses, parasites and bacteria can cause illness. *E. coli, Giardia* and *Cryptosporidium* can cause cramping, diarrhea, vomiting and other medical problems; *Legionella* causes respiratory distress. The Centers for Disease Control and Prevention (CDC) monitors waterborne disease outbreaks; in 2005 and 2006, only 20 community water systems (of the more than 155,000 systems) reported waterborne disease outbreaks. More than 600 people were affected in total, and three people died.

Many of the waterborne disease outbreaks you hear about in the news are related to recreational use—people who get sick from swimming or boating in contaminated waterways. Some illnesses are also contracted from untreated private water supplies.

12. Is tap water safer in one area of a community as compared with another?

Rarely. All the tap water from public water supplies must meet all federal, state or provincial requirements. In communities with a single source, everyone receives the same water. Some communities have more than one source, so different neighborhoods may have different quality water, but all the water must be safe to drink. The condition of the pipes and the flow patterns of water may be different in different areas, and this may cause some differences in water quality, although it usually does not affect water safety.

13. People are allowed to swim and go boating in our water supply reservoir. Should I worry about this?

Swimmers and boats do add some pollution, which is diluted by all the water in the lake or reservoir and doesn't amount to much. Also, because the water is thoroughly treated before it is delivered at the tap, any contamination is removed. Wildfires, litter and stormwater runoff can cause far more disturbance to the water quality in the supply reservoir than recreational pollution.

Still, some water districts do limit the activity on their water supply reservoirs to nonmotorized boats and prohibit swimming and other body-contact activities. Problems with methyl *tert*-butyl ether (MTBE), a gasoline additive, have been found in reservoirs that allow watercraft with two-stroke engines, such as Jet Skis, so some places have put strict restrictions on such craft. Prohibiting activities on a recreational water body can be an unpopular decision, so your water utility would need the support of its customers if such action were to be taken.

14. Should I use tap water for my baby's formula?

Because most tap water is pathogen-free, the majority of pediatricians consider tap water safe for mixing with baby formula. However, recent concerns have been raised by the Centers for Disease Control and Prevention (CDC) over infant formula made with tap water fluoridated with 0.7 milligrams per liter or higher. When infant formula concentrate is mixed with fluoridated water and *is the primary source of nutrition* for an infant, the chance of the child developing the faint white markings of very mild or mild enamel fluorosis may increase. Refer to the report, "Infant Formula and the Risk for Enamel Fluorosis," online at www.cdc.gov/fluoridation/safety/infant_formula.htm.

An American Academy of Pediatrics report also recommends regular water testing for nitrates in areas where only private well water is available. (Water suppliers regularly test for and control nitrates, so most publicly supplied supplies are safe.) The report, *Infant methemoglobinemia: the role of dietary nitrate,* outlines the dangers of nitrate-containing well water, especially for infants age three months and younger. Find the report at the National Guideline Clearinghouse, www.guideline.gov; search on methemoglobinemia. Check with your pediatrician if you have concerns about using your tap water for formula.

15. Is tap water suitable for use in a home kidney dialysis machine?

No, not without further treatment. The water used in a kidney dialysis machine comes in close contact with the patient's blood. Thus, the water needs to be much purer than ordinary drinking water. Aluminum, fluoride and chloramines may be added to water during the treatment process and are fine for drinking but are not acceptable in water used for kidney dialysis. Because chemicals such as these are critical for the safety of the water, the water must be further treated at the medical center immediately prior to use in a dialysis machine. Kidney dialysis centers are kept informed about water quality by water suppliers and are able to give advice to their patients on this matter.

16. Is it safe to take a drink from my garden hose?

It depends. To keep it flexible, a standard vinyl garden hose may contain unhealthy elements, such as lead, which can get into the water as it flows through the hose. These metals and chemicals are not good for animals either, so it's not a good idea to fill drinking containers for pets from a garden hose unless the water is allowed to run a while to flush out the chemicals.

However, some hoses do not contaminate the water because they are made with a "food-grade" plastic approved by the U.S. Food and Drug Administration. These hoses will be labeled "safe for drinking" and should be used on recreational vehicles when hooking up to a drinking water tap at a campsite. Remember, too, that the opening on any hose could be contaminated by chemicals or germs from lying on the ground or from prior use.

17. Is it safe to drink water from a drinking fountain?

Yes, usually. Some older, floor-standing water fountains are actually water coolers that contain lead-lined storage tanks, and there is a possibility that high amounts of lead could get into the water from these tanks. The U.S. Congress passed laws banning the use of lead in piping and in solder in 1986 and in storage tanks in 1988, and these regulations have improved the situation. One section of the 1988 act encourages schools and day-care centers to test their fountains for lead. Let the water run for a while before drinking from a fountain to minimize the risk. However, the storage tanks on most are about 1 quart (about 1 liter) in size, so complete flushing may take a while and the water you drink may not be as cold as that which has been held in the cooler.

18. Are in-store water dispensers safe?

In most cases, yes. Water from in-store dispensers often comes from the municipal water supply via a line that is linked directly to the store's main water pipes. Generally, the water is also treated in some way by the dispenser. The most common treatment is reverse osmosis. State health agencies usually inspect the dispensers on a regular basis, and stores must have policies about maintaining water safety between inspections to make sure the dispenser is carefully sanitized to prevent the possibility of contamination. Make sure the water jugs you are filling at the dispenser are sanitized between uses, as well.

19. Should I buy drinking water from a vending machine?

Buying water from a vending machine is a matter of personal choice. However, a 1998 report indicated that vending machine water can have higher bacteriological levels than tap water. Treatment within a vending machine can include reverse osmosis, activated carbon adsorption, and ultraviolet light disinfection. While such treatment can enhance the quality of many water supplies, vending machine treatment devices must be regularly maintained and water quality tested to provide satisfactory operation. If you use a machine in which you have confidence, be sure to put the water in very clean bottles, refrigerate it when you get home and treat it like a food.

20. Is there salt in my water?

Sodium or salt occurs naturally in drinking water. However, salt also finds its way into water from road de-icing, water treatment chemicals and ion-exchange water softeners. Sodium intake from tap water normally isn't a problem for the majority of people, but for those facing heart disease, hypertension, kidney disease, circulatory illness or a sodium-restricted diet, there are some legitimate concerns. Talk to your doctor if you have concerns about sodium intake.

MICROBIAL CONTAMINANTS

21. What types of living organisms do I need to be concerned about in water?

Living microscopic creatures occur naturally everywhere in our environment. *Pathogens* are germs—organisms

that cause disease. Common waterborne germs of concern are *bacteria* that cause botulism, typhoid, dysen-
tery, cholera and Legionnaires' disease; *viruses* that
cause hepatitis and polio; and *protozoa* that cause giardiasis and cryptosporidioses. The intestinal pathogenic
coliform *Escherichia coli* 0157:H7 (*E. coli* for short)
causes foodborne and waterborne disease outbreaks.

Utilities treat the water for all these pathogens and
more, and to determine if there are other microorganisms in the water that could cause harm to consumers,
they routinely test for harmless coliforms, common
organisms that live naturally in the intestines of
humans and aid in the functioning of the body.

22. How are germs kept out of my drinking water?

At most water treatment plants, a chemical *disinfectant*
is added that kills most living organisms in the water.
Chlorine gas and its liquid (sodium hypochlorite) and
solid (calcium hypochlorite) forms are the most common
disinfectants used in the United States and Canada.
The discovery of chlorine's effectiveness at killing cholera and typhoid germs has been heralded as one of the
most important health discoveries of the 20th century.
Other disinfectants include chloramines (a combination
of chlorine and ammonia), potassium permanganate,
ozone and ultraviolet (UV) light. Your water supplier
can tell you what disinfectant is used in your water.

Private water sources usually are not disinfected
and should be tested annually to uncover possible
contamination.

Some organisms, such as *Cryptosporidium*, are
chlorine resistant, and water suppliers that use surface
water use a multiple-barrier treatment process that
includes filtration and, at times, ozone or UV to remove
these organisms from your drinking water.

23. If I am concerned about my water, how can I kill the germs in it?

Using a timer, bring the water to a full boil on a stove or in a microwave oven and boil it for one minute. Because the boiling temperature of water goes down about 2 °F (1 °C) for each 1,000 feet (305 meters) you live above sea level, people living at high altitudes should increase the boiling time. For example, in Denver, Colo., which is more than 5,000 feet (1,525 meters) above sea level, boiling time should be increased to three minutes.

Treating water in this way should be done only in emergencies, because heating and boiling use a lot of energy, create a burn hazard and concentrate some chemicals in the drinking water. However the advantage of killing the pathogens in an emergency outweighs the slight disadvantage of concentrating the chemicals, which results in only a minor worsening of water quality. Always be careful with boiling water! Let it cool in a safe place.

24. Can flu viruses be spread in drinking water?

Treated tap water is not likely to transmit influenza viruses. Conventional disinfection processes that meet current drinking water treatment regulations provide a high degree of protection from viruses. Studies have shown that the H5N1 avian flu (bird flu) virus is killed by chlorine, and there are no known cases of humans getting the flu from exposure to drinking water.

25. What are *Cryptosporidium* and cryptosporidiosis?

Cryptosporidium—commonly called "crypto"—is a protozoan parasite that can live in the intestines of humans and animals (hosts). Outside of the hosts, the microbe is protected by a shell called an oocyst, so it is like a seed of a plant, very tough and long lasting. Once swallowed, the microbe emerges from its shell and infects the lining of the intestine. When this happens, some people get a disease called cryptosporidiosis. The usual time between swallowing this microbe and getting sick is 2 to 10 days. The major symptom is severe watery diarrhea, along with abdominal pain and loss of appetite, which last about 10 days to 2 weeks. Other symptoms can be nausea, vomiting, fever and headache. Although cryptosporidiosis is an unpleasant disease, it is not dangerous for people with normal immune systems. And not everyone who gets infected becomes ill.

Crypto gets into the waterways from the stools of infected humans and animals such as cattle, sheep and wild animals. Rain and melting snow run off over pastures and wildlands and transport the wastes into surface waters, rivers, lakes and streams. Wastewater treatment does not completely remove these microbes, so infected human wastes also contribute.

26. Are all water systems at risk from *Cryptosporidium*?

No. Protected groundwaters that are not mixed with surface water are usually free from these organisms.

27. Is drinking water the only source of *Cryptosporidium*?

No. There are many other sources. Unwashed fruits and vegetables, soil, swimming pools, recreational water, unfiltered stream water, day-care centers and nursing homes are also common sources. Remember, for all of these sources, the common factor is contamination from stools of infected humans or animals.

28. My water supplier found some of the microbes that cause cryptosporidiosis in the source water. Will I get sick if I drink water from the tap?

If you have a normal immune system, probably not. Because your water supplier was testing the source, it will be extra diligent during the treatment process to ensure that crypto is not making its way to the tap.

Also, the microbes detected by the utility might be harmless because the test doesn't differentiate between living and dead oocysts. Lots of water is sampled to look for the microbes, much more than would be in a glass of water, so finding a few in that large volume of water doesn't mean every glass of water has crypto in it.

The water supplier will also test the water after it's been treated, and if crypto is found then, the utility would issue a *boil-water order*. Even then, you would likely be okay. Although all people are presumed to be able to become "infected" from these microbes, some people have such strong immune systems that they will not get the disease. For example, estimates have shown that if 1 million people each swallowed one oocyst, only 5,000 would get infected, and only about two-thirds of these 5,000 would actually have the symptoms of the illness—a very small risk.

If you or a member of your family is at high risk of infection, however, you should take extra precautions. Infants, the elderly and immunocompromised people, such as cancer and transplant patients and those infected with HIV or who have AIDS, need to visit with their doctors and follow this advice from the CDC:

✓ Practice good hygiene (hand washing).
✓ Avoid water that might be contaminated.
✓ If you are unable to avoid using or drinking water that might be contaminated, make the water safe to drink by boiling it or filtering it.

✓ Avoid food that might be contaminated.
✓ Take extra care when traveling.
✓ Wash, peel or cook all vegetables.

For more information, go to the CDC's Web site, www.cdc.gov/crypto/, which has lots of valuable information about crypto prevention, detection and treatment.

29. Could my drinking water transmit the AIDS virus?

There is absolutely no evidence that AIDS can be transmitted through drinking water. There is no danger from drinking water for three reasons. First and most important, you can't get AIDS by drinking the virus; it must get into the blood directly. Second, the virus is very weak outside the body and rapidly becomes noninfectious. Finally, even if present in water sources, the virus is easily killed during the disinfection step of drinking water treatment.

CHEMICAL AND MINERAL CONTAMINANTS

30. Are chemicals found naturally in drinking water nontoxic?

Not necessarily. Many chemicals that occur in nature and seep into water supplies can be harmful to your health. A few examples are arsenic, radium, radon and selenium. Also, some nontoxic natural chemicals combine with other chemicals to produce harmful chemicals (by-products). Many naturally occurring chemicals are watched closely by your water supplier, which must test for numerous chemicals regulated by the USEPA.

31. What are organic chemicals? Are they dangerous?

Organic chemicals have mostly carbon atoms connected to hydrogen atoms (the element, not the gas). A common organic chemical in the home is sugar, so not all organic chemicals are toxic or cancer-causing. Food contains many beneficial organic chemicals essential to our life. You can't tell if something is an organic chemical just by looking at it. For example, table salt looks a lot like sugar but does not contain carbon and hydrogen, so salt is an inorganic chemical.

Dangerous organic chemicals are found in manmade products such as gasoline, cleaning fluid, pesticides, paint thinners and antifreeze. They are dangerous if they get into your drinking water because many are toxic, including cancer-causing chemicals called carcinogens.

USEPA regulates many of these chemicals, and when they are found at levels that exceed standards, water systems install specific treatment measures to remove these chemicals.

32. Where can I find out about how environmental exposures affect human health?

The Integrated Risk Information System (IRIS) database, prepared and maintained by USEPA, contains information about how environmental exposure to various chemicals can affect human health. Although initially developed for USEPA staff needing consistent information about these chemical substances, the information in IRIS is now available to the public at www. epa.gov/NCEA/iris/help_start.htm.

33. The movie "Erin Brockovich" focuses on chromium-6. What is that, and is it a threat to other water supplies?

Chromium-6 is an anticorrosion agent used in manufacturing, and it can be a health threat if it gets into water supplies and isn't treated. "Erin Brockovich" is based on a true story in which the Pacific Gas and Electric Company used chromium-6, also known as hexavalent chromium, in its compressor plant located near Hinkley, Calif. The chemical leachate from the plant contaminated the area's private groundwater wells. Hinkley residents reported many health problems, ranging from minor skin irritations to cancer and birth defects. The subsequent legal battle ended in an out-of-court settlement of $333 million for affected Hinkley residents.

USEPA and Health Canada regulate the amount of chromium-6 allowed in tap water, and water utilities regularly test for and control its presence.

34. Do hazardous wastes contaminate drinking water?

Yes. As rainwater seeps through a hazardous waste dump, it can carry hazardous chemicals to the groundwater. Some chemicals stick to dirt particles and don't reach the groundwater. Other chemicals, such as cleaning fluid, herbicides and gasoline, move rapidly down through the ground. Rain can also wash contaminants from a hazardous waste dump into surface waters, which can also seep into the groundwater and pollute it. This is one reason why there are such strict rules on covers and liners in the areas where hazardous wastes are disposed of.

Leaking underground gasoline tanks at gasoline stations and the improper disposal of chemicals (for example, dumping old radiator fluid, metal degreasers, paint thinners or paintbrush cleaners in the backyard) also may contaminate groundwater. Surface waters in some regions are contaminated by chemicals used for deicing roads in the winter. When it rains or the snow and ice melt, these chemicals wash into rivers, lakes and reservoirs.

To prevent this problem, some states post signs on roads that cross watersheds to inform the highway crews not to spread deicing chemicals in these areas. Improperly treated wastes from industrial plants may also pollute surface waters. Many water systems work hard to prevent such contamination, but if it occurs, these systems must treat their water to remove the chemicals.

Pharmaceuticals

35. Are there drugs in my water? How do they get there?

A variety of prescription and over-the-counter drugs (pharmaceuticals) and personal care products have been detected at very low levels in source water and some tap water in recent years. This includes small amounts of thousands of products people use every day, from aspirin to sunscreen and bug spray to cosmetics, deodorant, hair products, food supplements and caffeine. Collectively, the medicines and over-the-counter goods are called *pharmaceuticals and personal care products*, or PPCPs.

PPCPs enter the environment through the waste-water streams as they pass through people, or are washed off in a bath or shower. Other PPCPs are deliberately flushed down the toilet or poured into a sink. Perhaps the greatest contributor of drugs and hormones to the water supplies is livestock feedlots, where the hormones that are injected into pigs, beef, poultry and other animals are found in large quantities in waste ponds.

36. Should I be concerned about pharmaceuticals and personal care products in the water?

While no one wants to to drink someone else's medicine in their glass of water, anyone who takes medications or uses personal care products is exposed to them at much higher levels than can be found in drinking water. Also, improving technology means that increasingly tiny levels of PPCPs can be detected, so it appears that new contaminants of concern are constantly being found when in fact they've been in the water for a long time with no apparent ill effects.

This doesn't mean that there is no need to worry, because as the average age of the North American population has increased, more and more people are taking medications and using personal care products to feel and look better. Consequently, more such chemicals end up in the wastewater. Your water utility is already monitoring for harmful chemicals, and removing them if found, but more health-effects research is needed to determine the long-term consequences of PPCPs.

37. What are endocrine disrupting compounds (EDCs)?

EDCs are a particular group of chemicals that affect hormonal (i.e., endocrine) systems of animals. Hormones regulate reproduction and some behaviors, so some PPCPs, such as birth control pills, some pesticides and some antidepressants contain EDCs. Like PPCPs, EDCs can end up in the municipal and industrial wastewater treatment system either through direct discharge into the sewers or via stormwater runoff.

Aquatic life is particularly susceptible to the long-term effects of EDCs because it is constantly exposed to chemicals over multiple generations. More research is needed to determine long-term effects on humans, but thus far no one has linked drinking water to EDC-related problems in humans.

38. If flushing drugs down the toilet causes environmental problems, how should I dispose of unwanted medication?

Where available, take medications to a hazardous waste collection site or take-back program at a pharmacy or medical care facility. Before taking any controlled substance to a collection event, however, check to find out if the site is authorized to accept the material.

If that's not possible, keep the medicines in the original container. Remove the label or mark off your name and prescription number for safety. Add some water or soda to dissolve solid pills. Add something inedible like cat liter, dirt or cayenne pepper to liquids. Close the lid and secure with duct or packaging tape. Place the drug bottle inside an opaque container, like a coffee can or plastic laundry bottle. Tape that container closed. Place the container in the trash; do not put it in the recycle bin. Don't put it in with food that could be scavenged by humans, pets or wildlife.

The USEPA and U.S. Food and Drug Administration (FDA) do recommend flushing certain controlled drugs that are particularly powerful and/or addictive, including oxycodone, morphine, fentanyl patches and gatifloxacin. Be sure to read the disposal information provided with the drug for guidance, or check out FDA's Web site, www.fda.gov/ForConsumers/default.htm.

Arsenic

39. What is arsenic and how does it get in my drinking water?

Arsenic is an odorless and tasteless element that can occur naturally in the Earth's soils, rocks and minerals. Arsenic can also enter drinking water supplies from agricultural and industrial activities. Industrial arsenic in the United States is primarily used as a wood preservative, but arsenic is used in paints, dyes, metals, drugs, soaps and semiconductors. Certain fertilizers and animal feeding operations can contribute to arsenic contamination as well, along with copper smelting, mining and coal burning.

Groundwater flowing through natural arsenic deposits can dissolve the mineral, resulting in elevated amounts of arsenic in well water. Other sources of groundwater contamination include runoff or seepage from minefields, feedlots and industrial waste sites.

40. What are the health effects of arsenic exposure?

High doses of arsenic or chronic exposure over a long period of time can cause thickening and discoloration of the skin, stomach pain, nausea, vomiting, diarrhea, numbness in hands and feet, partial paralysis, diabetes and blindness. Arsenic poisoning also has been linked to cancer of the bladder, lungs, skin, kidney, nasal passages, liver and prostate.

41. Is arsenic regulated in drinking water?

The USEPA standard and Health Canada guideline for arsenic in drinking water are the same: 10 micrograms (parts per billion) per liter. This is about one thousand times lower than the amount of arsenic it would take to cause health effects.

If you use a private well, you should consider having it tested for arsenic. Because arsenic has no taste or smell in drinking water, the only way to determine whether it is in your well water is by having a water sample tested by a certified state or commercial laboratory. A list of certified laboratories should be available from your local or state health department. Some certified labs may also be listed in the yellow pages of your telephone book.

42. What home treatment systems best remove arsenic?

Ion exchange, reverse osmosis and distillation point-of-use (POU) systems are effective at reducing arsenic levels in most home water supplies, but USEPA does not endorse POU units for small community water systems to treat for arsenic; this means there are issues with using this technology for this purpose. Before installing any treatment system, thoroughly investigate the system's ability to remove arsenic.

USEPA has an environmental technology verification program that has tested numerous products for their effectiveness in removing arsenic and other inorganic compounds. For reports on individual products, go to www.epa.gov/nrmrl/std/etv/vt-dws.html.

Atrazine

43. What is atrazine?

Atrazine is an herbicide widely used in the United States and Canada to control broadleaf and grassy weeds, particularly in corn and soybean crops.

44. Can atrazine get into the water and affect human health?

Atrazine is a known water contaminant. It typically enters surface waters by running off crop fields after late spring or early summer rainstorms. The chemical also enters the water from spillage or accidental discharge during production, packaging, storage and waste disposal. Significant contamination of groundwater aquifers has occurred in areas where atrazine is used extensively.

Atrazine has been associated with causing hormone imbalances in laboratory animals, possibly disrupting reproductive and developmental processes on a short-term basis. USEPA has determined that atrazine is not likely to be carcinogenic to humans, but in Canada, atrazine has been classified as possibly carcinogenic to humans, based on studies in rats dosed with atrazine that resulted in an increase in mammary and uterine tumors, leukemia and lymphomas.

45. Is atrazine regulated?

USEPA has established a maximum contaminant level (MCL) of 3 parts per billion (3 micrograms per liter) for atrazine, and Canada has an Interim Maximum Acceptable Concentration of 5 micrograms per liter of drinking water. Both levels are being reviewed in light of new health-effects data that has been generated through atrazine's re-registration process at USEPA's Office of Pesticide Programs. Meanwhile, water utilities are required to test for atrazine and treat the water to eliminate it or reduce it to safe levels at the tap.

If atrazine is found in your public water supplies, information about it will be included in your utility's annual water quality report, which is delivered to every customer each year and may also available online.

Lead and Copper

46. How does lead get into drinking water?

Where lead is present in home plumbing and in soldered connections, it may dissolve into the water while the water is not moving, generally overnight or during other times when the water supply is not used for several hours. Faucets with brass or bronze internal parts may also contribute to lead at the tap.

The first water that is drawn from a faucet after long periods of nonuse may contain lead. Naturally soft low-mineral-content water picks up more lead than hard water because water with low amounts of dissolved minerals tends to dissolve metal pipes, and hard water has a tendency to lay down a scale layer on the inside of pipes that keeps the lead (or copper) from leaching. This is why water utilities may add phosphates or other non-corrosive minerals to the water at the treatment plant.

USEPA regulates lead in drinking water, which requires corrective action by the supplier if high amounts are found at the tap. Water suppliers must offer to sample the tap water of any customer who requests it, but the supplier is not required to pay for collecting or analyzing the sample, nor is the system required to collect and analyze the sample itself.

Canada banned the use of lead pipe and lead-based solder in 1990 and currently has a guideline of 0.01 milligrams of lead per liter.

In the United States, lead is banned in pipes and solder, and the voluntary standard for lead in faucets has caused the plumbing industry to change manufacturing methods to reduce lead levels. An NSF mark on a faucet means that NSF International has tested the faucet and it meets the standard. Consumers should insist that no plumbing installations or repairs be done with fixtures containing lead or lead-based solder.

47. How can I get lead or copper out of my drinking water?

If testing has indicated a problem, if your water is corrosive or if you have rusty water or blue-green stains in your sink, take the following precautions.

Whenever water has not been used for a long period of time—overnight or during the day if no one is home—let cold water run from the faucet for about two minutes or until you get colder water before using any water for drinking or cooking. Just how long it takes for fresh water from street pipes to arrive at the faucet depends on your specific location, water pressure, whether you live in a single-family home or an apartment and so forth. Save the first-draw water for other purposes, such as plant watering. Running the water for two minutes may not flush out all the lead that got into the water while it was sitting in your plumbing, but it should dilute it to acceptable levels.

Home treatment equipment that contain lead-removing filters, reverse osmosis systems and distillation units remove lead dissolved in water. Check to see whether the performance of these products has been tested for lead reduction by independent testing and certifying organizations following the methods contained in the appropriate ANSI/NSF "Drinking Water Treatment Unit" standard. More information about home drinking water units can be found on NSF International's Web site, www.nsf.org/certified/dwtu.

48. What are the health effects of drinking water with lead in it?

Lead may cause a range of health effects including delays in physical or mental development, particularly in children six years old and under, because this is when the brain is developing. For adults, lead can cause kidney problems or high blood pressure, and a recent study has linked lifetime exposure to lead from drinking water pipes to an increased risk of cataract development in men.

The primary source of lead exposure for most children is ingesting lead-based paint chips and inhaling paint dust in older homes. USEPA estimates that 10 to 20 percent of human exposure to lead may come from lead in drinking water. Infants who consume mostly mixed formula can receive 40 to 60 percent of their exposure to lead from drinking water. For more information about lead, visit the USEPA's lead information Web site, www.epa.gov/safewater/lead.

In recent years, USEPA has placed stricter requirements on water utilities for lead monitoring, testing and treatment, lead service line management, and customer awareness through the Drinking Water Lead Reduction Plan. The plan also aims to reduce lead in drinking water by addressing lead in tap water in schools and child-care facilities.

Nitrates and Pesticides

49. How do nitrates and pesticides get into my drinking water? What health problems do they cause?

Nitrates and pesticides come from fertilizers and pesticides used on farms and on home gardens and lawns. Septic-tank drain fields and wastes from animal feedlots are other major sources of nitrates. Rainwater takes these chemicals with it and contaminates groundwater and waterways. As rain moves through the soil, microbes in the soil change the ammonia in the fertilizer and drainage from septic tanks and feedlots into nitrates.

USEPA and Health Canada have set a limit for nitrates because high levels are associated with a rare blood condition in infants, commonly called *blue-baby syndrome*. The name comes from the baby's skin, which

turns bluish after the baby drinks water containing nitrates because the chemical interferes with the oxygen-
carrying capability of hemoglobin in the blood.

If your water supply comes from a private well,
you may want to hire a local water treatment dealer to
have your water tested and install proper equipment,
if necessary, to remove nitrates and pesticides.

Perchlorate

50. What is perchlorate?

Perchlorate is both a naturally occurring and man-made
chemical. Perchlorate is used to manufacture fireworks,
explosives, flares and rocket propellant. It forms when
chloride and oxygen join together and is commonly
found as part of other substances, such as ammonium
perchlorate, potassium perchlorate, sodium perchlorate,
and perchloric acid. Perchlorate dissolves easily and
moves quickly in underground water and surface water.
It breaks down very slowly in the environment.

Perchlorate has been has been found in the water
supplies of 26 states and two territories, particularly
California, Nevada and Utah, representing just over
4 percent of public water systems nationally.

51. Is perchlorate regulated?

USEPA and water suppliers have been monitoring perchlorate in water supplies for more than a decade, and
the chemical is on USEPA's Contaminant Candidate List.
Because of disagreements about the interpretation of the
data, USEPA chose not to issue a federal drinking water
standard for perchlorate in 2008; however, the agency
is revisiting the issue and expects to pass perchlorate
regulations in the future.

Meanwhile, USEPA has issued an Interim Drinking Water Health Advisory level of 15 parts per billion (15 micrograms per liter) of perchlorate in drinking water. This is even lower than the advisory "action level" of 18 ppb of perchlorate in drinking water adopted by the California Department of Health Services.

52. How can perchlorate be removed if it gets in my drinking water?

Some reverse osmosis home-treatment units have been certified by NSF International to remove perchlorate from levels as high as 130 µg/L to 4 µg/L or less in drinking water. However, before installing a home treatment unit, contact the manufacturer to determine if the unit can remove perchlorate from your water supply.

Radon and Radium

53. What is radon and is it harmful in drinking water?

Radon is a radioactive gas that is dissolved in some groundwaters. It is formed when radium or uranium decays naturally. When inhaled over long periods of time, radon can cause cancer. Experts also think radon has some harmful effects when consumed. When water that contains radon is exposed to the air, the radon is released and some remains in the water. However, radon in drinking water generally contributes a very small part (about 1 to 2 percent) of total radon exposure from indoor air.

USEPA proposed regulations to reduce the public health risks associated with radon in air and water in late 1999, but final action on the proposal was still pending as of the summer of 2009. Water suppliers do have treatment methods available for radon removal when and if it is required.

For more information about radon in the air, go to EPA's radon Web site at www.epa.gov/radon.

54. My private well has high levels of radon. How can I make sure my drinking water is safe?

The most effective treatment is to remove radon from the water right before it enters your home with a point-of-entry (POE) treatment unit. Two kinds of POE devices remove radon from water: granular activated carbon (GAC) filters and aeration. Before you do this, however, discuss the process and consequences with local authorities; after collecting radon, GAC becomes a hazardous waste, and aeration puts radon into the air. There may be regulations in your community that control what can be done with excess radon.

55. What is radium and how is it treated?

Radium is a radioactive element that occurs naturally in some soils and some groundwaters, along with radon. Radium is regulated by USEPA, and water utilities monitor for it and remove it if found in water sources.

Small water systems and private well owners can use ion exchange and reverse osmosis point-of-use devices to remove radium. Water softeners also effectively remove radium, but they do not remove radon. In fact, the radium that is trapped by the water softener may create some radon. Discuss this issue with your water supplier.

WATER TREATMENT

56. Are all chemicals in my drinking water bad for me?

No. Some chemicals are good for you, and some minerals are accepted by most to be beneficial in drinking water, such as fluoride at proper levels. In addition, many chemicals are necessary to improve the quality of the water, and these should have no effect on your health.

Chlorine

57. Is water with chlorine in it safe to drink?

Yes. Chlorine is added during the water treatment process to kill microbial contaminants and a residual amount is left in to keep the water safe in the distribution system. USEPA limits the amount of chlorine remaining in the water at the tap to safe drinking levels, and many tests have shown that this amount is safe to drink, although some people object to the taste or smell. If this is a problem for you, chill a pitcher of tap water in the refrigerator, or let the water stand in the glass for a short while before drinking, and the taste and smell should dissipate.

58. Is there a link between chlorine and cancer?

During water treatment, chlorine can combine with certain naturally occurring chemicals to form compounds that may cause cancer. This can happen with chemicals released by certain algae blooms in water supply reservoirs, or from the decomposition of any plant material in the water. USEPA calls the resulting compounds *disinfection by-products* (DBPs) and has set strict limits on the DBP level allowed in treated water. DBPs regulated by USEPA include a group of four chemicals with the general name trihalomethanes (THMs). Another group includes five haloacetic acids (HAA5). If these reaction products are in your tap water at harmful levels, you will be notified by your supplier. If you're concerned, call your supplier and ask whether the total THM or HAA5 level in your water is okay.

Remember, disinfection is an important part of the treatment process, and disinfectant levels must remain adequate to kill the germs found in water. Any harmful effects to humans from DBPs are very small and difficult to measure in comparison with the risks associated with inadequate disinfection.

59. Should I be concerned about the chlorine in the water I use for bathing or showering?

No. Chlorine does not absorb into the skin and get into your body, and the amount of chlorine in the water is too low to harm the skin itself. The levels are also too low to cause problems breathing when the chlorine is released from the water into the air during a shower.

Some people may be allergic to chlorine and related compounds, however, and this is particularly problematic in and near swimming pools, where the amount of chlorine is much greater than in tap water. Related studies have shown that asthmatic reactions of some swimmers and lifeguards are related to the chlorine by-products, not the chlorine itself.

60. What else is used to kill the germs in water?

There are several alternative disinfectants, but each has its own limitations. *Chloramines* are a combination of chlorine and ammonia, and they form very few disinfection by-products. But chloramines are not as strong as chlorine, so they are generally used only as a supplemental disinfectant to prevent contamination in the distribution system.

Ozone use is increasing in the United States, but water plants must add the equipment to handle the highly corrosive and toxic ozone gas. Ozone is highly effective and has no taste, but it dissipates quickly and doesn't provide protection in the distribution system, so chloramines or chlorine must be added to product the water all the way to the tap. This is also true for *ultraviolet* (UV) light. UV does not form DBPs and is an effective disinfectant in very clear water, but very high doses of UV radiation are needed to kill large protozoa such as *Cryptosporidium* and *Giardia*.

Chlorine dioxide is highly efficient, particularly in cold water, but it is highly volatile and hard to store.

Potassium permanganate ($KMnO_4$) is used to supplement chlorine disinfectants and thus reduces the potential for disinfection by-products to form. Otherwise, the primary use for $KMnO_4$ is to control taste and odors, remove color and remove iron and manganese.

Aluminum

61. I hear aluminum is used to treat drinking water. Is this a problem? Does it cause Alzheimer's disease?

Aluminum-containing chemicals, called *alum* or *aluminum sulfate*, are used to treat most surface waters. These chemicals trap dirt and then form large particles in the water that settle out; thus, very little aluminum stays in the water. Aluminum is a natural chemical that occurs in many foods, including tea; even if you live in areas where the level of aluminum in drinking water is much above average, your intake from food would be about 20 times your intake from drinking water.

Preliminary studies conducted by the Water Research Foundation in conjunction with health professionals shows that aluminum is not a primary cause of Alzheimer's disease. However, the debate and research continues on whether aluminum contributes to the progression of the illness and whether the contribution of alum in drinking water is more problematic than aluminum from other sources. An executive summary of the report can be found at www.water-researchfoundation.org/research/topicsandprojects/execSum/722.aspx.

USEPA considers an acceptable range of aluminum drinking water to be between 0.05 milligrams per liter (mg/L) and 0.20 mg/L. This is a recommendation, not an enforceable standard.

Health Canada also does not have a health-based standard for aluminum, but recommends that plants reduce residual aluminum levels in treated water to the lowest extent possible.

Fluoride

62. Is the fluoride in my drinking water safe?

Yes. Fluoride helps build strong, healthy teeth that resist decay. When added or naturally present in the correct amounts, fluoride in drinking water has greatly improved the dental health of American and Canadian consumers since it was first introduced in the mid-1940s. The CDC, American Dental Assocation, American Medical Association, American Water Works Association, U.S. Surgeon General and numerous other professional groups endorse fluoridation of community water supplies because of the public health benefit it provides. Water fluoridation has been recognized by the

CDC as one of 10 Great Public Health Interventions of the 20th Century.

Fluoridation even saves money: CDC estimates that for every dollar spent on fluoridating water supplies, $68 is saved in dental costs.

More information on fluoride and its benefits can be found on the CDC Web site: www.cdc.gov/fluoridation.

63. Will I lose the benefits of fluoride in my drinking water if I install a home treatment device or drink bottled water?

Certain types of home treatment devices will remove 85 to more than 95 percent of all the minerals in water, including fluoride. These are reverse osmosis, distillation units and deionization units (not water softeners—they leave fluoride in the water). If you use one of these types of devices, consult with your dentist about fluoride and possibly your doctor.

The situation with bottled water is less clear. Many bottled waters contain very little fluoride, although a few contain adequate amounts. Bottlers are not required to list the fluoride content in a bottle of water, unless it is added at bottling. If the bottled water contains greater than 0.6 milligrams per liter (mg/L) and up to 1.0 mg/L, the label may contain the statement, "Drinking fluoridated water may reduce the risk of tooth decay." Otherwise, you'll have to contact the bottler to determine how much fluoride, if any, is in the water.

TRAVEL

64. If I travel overseas, is the tap water safe to drink?

Besides the United States and Canada, the water generally is safe to drink in many industrialized countries.

Many developing countries, however, have inadequate sanitation and water supply infrastructure, so precautions should be taken when visiting these areas.

The World Health Organization (WHO) has a searchable database of information about water quality, sanitation and multiple other health statistics for more than 190 countries available on the Web: www.who.int/whosis/en.

The CDC recommends that international travelers take a number of simple precautions before and during travel to avoid potential health problems:

✓ Contact their physicians, local health departments or agencies that advise international travelers at least 4 to 6 weeks before departure to schedule an appointment to receive current health information on the countries they plan to visit.
✓ Obtain vaccinations and prophylactic medications as indicated.
✓ Address any special needs.

Health Canada also has an active travel advisory site: www.hc-sc.gc.ca/hl-vs/travel-voyage/index-eng.php.

65. Is there something I can take with me when I travel to purify water?

Several portable mechanical water purifiers claim to produce germ-free water, one glass at a time. These devices frequently combine a filter and a chemical disinfectant and are available at outdoor gear stores. Some portable purification devices on the market use ultraviolet light or other methods that do not involve chemicals and are not required to be registered by USEPA.

Because a traveler cannot determine the proper conditions required to make sure that ultraviolet light will kill *Cryptosporidium* and *Giardia*, any device that uses only ultraviolet light should be used with caution.

Some hikers and travelers also use purifying tablets or iodine, but these usually add an odd taste to the water and do not protect against *Giardia* or *Cryptosporidium*.

66. When I travel to a different place in this country, sometimes I have an upset stomach for a couple of days. Is this because something is wrong with the water?

If the water comes from a public water system, your stomach problems are likely not a result of contamination problems in the drinking water. However, waters with a high mineral content, particularly sulfate, may have a temporary laxative effect if your body is not accustomed to the water. Therefore, the change in mineral content from place to place sometimes does bother travelers for a short time until the body readjusts.

67. I've heard that some countries use solar rays to disinfect their drinking water. Is that true?

Solar disinfection has proved useful in developing countries, where access to safe water is limited. Solar disinfection relies solely on the heat and ultraviolet radiation of the sun to make water stored in transparent containers drinkable.

The practice is used at the household level under the responsibility of the user, and its success rate depends on numerous factors, including the intensity of sunlight, geographic location, cloud cover, time of day, type of bacteria being exposed, containers in which the contaminated water is kept during exposure and clarity and depth of the water.

68. How is water quality protected on airplanes?

Because water on U.S. flights comes from public water systems, the water quality on aircraft is governed by USEPA under the Aircraft Drinking Water Rule. The rule is specifically aimed at safeguarding airline supplies by requiring airlines to have their airplane water systems inspected at least once every five years and to fix any significant problems.

The rule was proposed after an investigation revealed contaminated water supplies in the aircraft of several major airlines. The airline industry has since implemented new aircraft water testing and disinfection protocols to protect the traveling public. Health Canada found a similar situation in 2006, and now works with airlines to improve onboard water quality. The protocols emphasize preventive measures, adequate monitoring and sound maintenance practices such as flushing and disinfection of aircraft water systems.

For more information on this topic, go to USEPA's Web site: www.epa.gov/safewater/airlinewater/index2. html. In Canada, go to www.hc-sc.gc.ca/hl-vs/travel-voyage/general/airplane-aeronefs-eng.php.

69. We sometimes hear about waterborne illness outbreaks on cruise ships. How is water quality protected on cruise ships?

Unlike outbreaks in a land community, cruise ships provide a controlled environment in which it is easy to track the spread of disease, and yes, most infectious disease outbreaks on ships are water- or food-borne, caused by unsanitary practices (lack of hand washing) or other conditions.

That said, most cruise ships operating out of U.S. ports participate in the voluntary Vessel Sanitation Program run by the CDC. CDC personnel conduct unannounced inspections of the cruise ships twice a year and rate the cleanliness of food and water on a 100-point scale. Drinking water items make up 27 of the 100 points. Most ships take water from the shore, disinfect it and put it in tanks on the ship (this is called *bunkering*). Some ships do, however, make drinking water from sea water. Whatever the source, the water then must be disinfected again before going to the passengers, and records must be kept to show that the disinfection was up to standards.

For more information about a specific ship or an outbreak, visit www.cdc.gov/nceh/vsp. Health Canada also has a list at www.hc-sc.gc.ca/hl-vs/travel-voyage/general/index-eng.php.

70. How is drinking water on cruise ships treated?

Cruise ships used to bring water from shore, but with the biggest ships using more than 250,000 gallons (947,000 liters) of fresh water a day, this is impractical. Most of these liners have on-board desalination units, either employing reverse osmosis or flash evaporators that boil seawater and recondense the steam vapor, to produce purified water from the ocean water. This water then goes through a remineralization process to add back minerals that give the water a fresh flavor, followed by disinfection to kill any microbial contaminants.

The wastewater also goes through an extensive treatment process before being discharged.

71. Do I need more water if I'm moving or traveling to a higher altitude?

Dehydration is a valid concern when adjusting to a higher altitude because the altitude can trigger rapid breathing and more frequent urination. Each person's reaction will differ, depending in part on how much of an altitude change is being made. The basic six to eight 8-ounce glasses of water per day should be the minimum intake, more if you are involved in strenuous activities. If you have any special health concerns, it is always a good idea to consult with your doctor before traveling to higher altitudes.

72. Is it okay for campers, hikers and backpackers to drink water from remote streams?

No. These streams often contain the pathogenic protozoa *Giardia* or *Cryptosporidium* and other organisms that can cause disease.

73. What can campers, hikers and backpackers do to treat stream water to make it safe to drink?

In addition to or in place of purchasing a water purifier as discussed in a previous question, any water that looks good enough to drink can be made microbiologically safe by boiling. One minute of vigorous boiling at sea level or three minutes at high elevations will kill all germs, disease-producing bacteria, viruses and protozoan cysts.

Water disinfection tablets, available at drugstores and camping supply outlets, can be put into a glass of clear water. They take about five minutes to dissolve and release the disinfectant. These tablets will not protect against *Giardia* and *Cryptosporidium* (the primary concerns in most areas) or parasitic worms, however, and will not work well in cloudy or colored water. Thus, their use alone is discouraged.

Taste, Odor and Appearance

Because of you I notice the taste of water,
a luxury I might otherwise have missed.
— Adrienne Rich, *Like This Together,* 1963

74. Can I tell if my drinking water is okay by just looking at it, tasting it or smelling it?

Ideal drinking water is crystal clear, has no distinct smell, no lingering aftertaste and no mouth-feel sensation such as a drying feeling or tongue coating. But water can have an odor, taste or color for many reasons that do not affect the safety of the water, including natural minerals, such as sulphur, or organic compounds, such as tannins, that sometimes can be found in water sources.

Contrarily, water that looks, smells and tastes good may also contain undetected microorganisms or chemical contaminants that can cause disease. For example, clear, cold stream water sometimes contains *Giardia lamblia*, a parasite that causes intestinal illness.

Chemicals are a different matter. Chemicals cannot be seen in water, but many do impart tastes or odors. Fortunately, the U.S. Environmental Protection Agency's (USEPA) health limits and proposed limits on regulated chemicals are much higher than the amounts that cause tastes and odors. This means that even if you taste or smell a chemical odor, the water may still be safe to drink. But you should still report any problem like this to your water supplier.

USEPA also sets standards for water's aesthetic qualities called Secondary Maximum Contaminant Levels (SMCLs). These are not enforceable, but most public utilities abide by SMCLs to control overt smells, odors or colors in tap water. Your drinking water should be clear and nearly taste-free and odor-free. If your water quality changes, report it to your water supplier immediately as it could signal a problem.

75. Why does my drinking water taste or smell "funny"? Will this smelly water make me sick?

Our senses of taste and smell do play an important role in protecting us against unsafe foods and beverages, but serious illness seldom results from taste and odor (T&O) problems with your water. Some contaminants that could affect your health can be tasted in drinking water (e.g., gasoline, oil, and ethylene glycol from backflow events), and some harmful contaminants (such as lead) cannot be tasted. T&O issues occur from a variety of reasons, including:

Chlorine

People with highly astute senses may notice a flavor or odor in water from the chlorine that is added by your water utility to kill bacteria and viruses. Chlorinated water may contain reaction by-products (known in the industry as *disinfection by-products*) which, in excess, can be a concern to consumers. These products cause no taste or odor at detectable levels and are restricted by USEPA rules, which means your water utility takes steps to control the amount of by-products, as well as the level of chlorine, in your water. Nonetheless, many utilities are required by USEPA to make sure that every tap in their systems has a detectable level of chlorine in the water to protect the water once it is in the distribution system.

Rotten-Egg Odor

A rotten-egg odor in some groundwater is caused by a nontoxic (in small amounts), smelly chemical—hydrogen sulfide—dissolved in the water. This is a common issue with many private wells, and anyone who has visited natural hot springs knows the smell. A rotten-egg odor may be a sewer smell; if you are uncertain, have a plumber check it out.

Earthy, Musty Odors

As some algae, bacteria and tiny fungi grow in water, they give off nontoxic, odorous chemicals that can cause unpleasant odors in drinking water. Different microorganisms can produce different tastes and odors; the two most common ones are identified as "earthy" and "musty." Another possible culprit of this T&O issue is stale water that hasn't "turned over" because it has been standing in service lines or little-used lines in a new housing development.

Metallic Taste

Metallic tastes can come from copper that has dissolved from copper pipe and from iron from rusting iron pipes. Iron can also be found in well waters at levels that can impart a metallic taste. Higher levels of copper can cause short-term health problems like diarrhea and cramping. Iron generally has no effect on health at the levels that could be consumed through water.

Sewer Smell

Sewer-like smells in your water can be caused by a cross-connection with a wastewater collection system component and could be a serious problem. Have a certified plumber inspect your plumbing for proper connections and valves to prevent this from happening.

Flat-tasting Water

Water usually tastes flat because it lacks oxygen or minerals. Water that has been treated with reverse osmosis and distilled water can taste flat to many people, as can stored or boiled water. In fact, some bottled water suppliers add minerals to their water to prevent it from tasting flat.

You should report any sudden change in taste or smell in your drinking water to your water supplier.

76. What can I do if my drinking water tastes "funny"?

Ways to remedy the problem of "funny" or undesirable-tasting water include:

✓ Store drinking water in a closed glass container in the refrigerator, because cold drinking water has less flavor than warm drinking water. Although some plastic bottles are okay for storing drinking water in the refrigerator, other types of plastic will cause a taste in water.

✓ Flat-tasting water, usually from storage, will taste better if you pour it back and forth between two clean containers before drinking it. This will put oxygen from the air back into the water and remove the flat taste.

✓ Add 1 or 2 teaspoons of lemon juice to refrigerated drinking water.

If you draw water from a private well or your water consistently has T&O problems, consider installing a point-of-use (POU) water treatment product that has been tested by an independent organization following the method in the Taste and Odor Reduction section of ANSI/NSF Standard 42, Drinking Water Treatment Units – Aesthetic Effects.

POU devices often contain activated carbon (sometimes mistakenly called activated charcoal or just charcoal), which can remove many taste- and odor-causing chemicals, including chlorine. Some devices use reverse osmosis (using a membrane) or distillation instead of activated carbon. POU devices can be tap mounted or located under the sink. If you wish to treat all of the water entering your home, you'll need to install a point-of-entry (POE) device at the main service line coming into your house. Filtering devices can cost a few dollars up to $1,000 and generally do a good job in treating T&O problems. If you plan on storing water from a home filtering device, store it like you would food, using clean, airtight containers and refrigerating.

If the problem is a rotten-egg odor, consider installing home treatment equipment that will remove hydrogen sulfide. First, however, make sure the odor is not coming from the hot water heater or the drains in the sink.

If you have a water softener that treats both hot and cold water, the chlorine will react with the softening materials inside the softener, and the chlorine will be removed.

Contact your water supplier with your concerns if you receive your water from a public or investor-owned water system. Water supply system employees are usually aware of current T&O problems in their systems and are ready to answer customer questions. Your call may alert them to an emerging problem, so report any unusual taste or odor to your water supplier.

77. Why do the ice cubes from my freezer used to cool my water make the water taste funny?

This is a common complaint that has no single, simple explanation. Many items in a refrigerator and freezer can give off odors that are absorbed by the ice. Freezers usually contain packaging materials, food, and metal or plastic ice cube trays. If you have an automatic ice-maker, harmless bacteria can grow in the water feed line and cause odors. Smelly substances in use near a freezer may be absorbed into the ice. "Freezer smell" can sometimes be noticed in empty metal ice cube trays. Though annoying, these "off flavors" are not harmful and can sometimes be lessened by cleaning and defrosting your freezer and ice cube trays more often. To help prevent odors in the freezer altogether, put an open box of an odor absorbent, such as sodium bicarbonate (baking soda), in the freezer and refrigerator units.

78. White stuff appears in the glass as ice cubes from my freezer melt. What is the white stuff and where does it come from?

Ice cubes freeze first on the outside, so the center of the cube is the last to freeze. As an ice cube freezes, the dissolved minerals in the water are pushed to the center and are trapped there. Near the end of the freezing, when not much water remains in the center of the cube, the minerals become very concentrated and form the white stuff, which is technically called *precipitate*. This precipitate causes the white stuff in your glass and is not toxic.

Some commercial ice cubes are "cored" after they freeze to remove the precipitate. Having posts in your ice cube tray doesn't help, as the precipitate must actually be removed by coring. And there is a new fad being reported: some high-end restaurants are serving expensive designer ice cubes in their cocktails; according to one manufacturer, these cubes are ultrapurified to remove the minerals and chemicals that the manufacturer claims affect "the flavor and integrity of the drink or the ice itself." [1]

79. Why does drinking water often look cloudy when first drawn from a faucet?

The cloudy water could be caused by tiny air bubbles in the water similar to the gas bubbles in carbonated soft drinks. After a while, the bubbles rise to the top and are gone. This type of cloudiness occurs more often in the winter, when the drinking water is cold and holds more air, and then is drawn into the house and heated up in the plumbing.

Another cause of cloudiness in cold water comes from calcium. In certain waters, the nontoxic insoluble chemical calcium carbonate will precipitate when it is cold. Because it is white, this precipitate can cause the water to look cloudy, but the particles will settle to the bottom of the glass (usually in about 30 minutes), in contrast to the air bubbles previously discussed that rise to the top of the water fairly quickly. Water containing calcium carbonate precipitate is perfectly safe to drink or use for cooking, though it may be unappealing to the eye.

80. My drinking water is reddish or brown. What causes this?

Most reddish-brown color found in tap water is nontoxic, but it can be a nuisance. It can stain fixtures and clothing in the washer, and it looks bad. There are several possible causes.

---

1   Amter, Charlie. Daily Dish. Los Angeles Times. Jan. 19, 2009

Natural organic matter produces tannins in many well waters. Known as humic acids, they are the dark brown constituents caused by decaying organic substances in soil, and can also be found in peat, coal, many upland streams and ocean water.

Tannins also come from decaying vegetation that leaches its color into the source water, similar to the way water changes color after tea leaves are added to it. Typically this color must be removed by the treatment plant, although some point-of-entry devices on the market claim to filter out color.

Iron, which is common in nature, may be dissolved in your drinking water. When iron is dissolved in groundwater, it is colorless, but when it combines with air as water is drawn from your faucet or elsewhere in the system, the iron turns reddish-brown. This can happen before your eyes after a glass of tap water is drawn. If you notice the water changing from colorless to brown, you may want to consider buying an iron-removal unit for your home.

Drinking water pipes—in the street, leading to your home, or in your home—may be rusting, creating rusty-brown water. Galvanized pipes could be one source. Also, your hot water tank may be rusting. Corrosive water can cause this type of problem. If you are having trouble and your neighbors are not, then your own pipes or water heater probably cause the problem. When your plumbing is rusting, lead and copper may be getting into your drinking water as well. This is important, so call your local water supplier or have a plumber check your plumbing. You might have an electrical circuit grounded improperly to your water pipes, which induces corrosion of the pipes. Letting the water run a while will often clear the water (save the rusty water for plants). To avoid problems with lead and copper, USEPA has established regulations that limit the amount of lead and copper allowed in your water, and all the water suppliers by law have to make sure that drinking water is not corrosive.

81. My drinking water is dark in color, nearly black. What causes this?

When manganese, a mineral that frequently occurs in nature, dissolves in groundwater, it is colorless. (Human activities, particularly manufacturing and agriculture, are also responsible for manganese found in water in some areas.) When manganese combines with the chlorine in the water, it turns black. To prevent black water problems, USEPA has established a recommended limit (not required, just recommended) for manganese in drinking water, and many water treatment plants have oxidation processes to eliminate the manganese before it reaches your tap. If you have blackish water, you may want to consider installing a point-of-use filter. Ion exchange water softeners may help. You should also report your problem to your water supplier.

Manganese found in water is generally not considered a health threat, and a variety of foods and beverages also contain natural manganese. Small amounts of both manganese and iron are essential to a healthy diet.

In the Home

Water, like religion and ideology, has the power to move millions of people. Since the very birth of human civilization, people have moved to settle close to it. People move when there is too little of it. People move when there is too much of it. People journey down it. People write, sing and dance about it. People fight over it. And all people, everywhere and every day, need it.
— Mikhail Gorbachev, quoted in *Peter Swanson's Water: The Drop of Life,* 2001

82. How much water does one person use each day?

Most people in the United States drink about a half-gallon of water per day. The U.S. average is nearly 160 gallons (606 liters) used each day per person for all home uses. Total per-home water use includes lawn irrigation and washing clothes and dishes, as well as use in toilets, showers and baths. Of this, the amount used for cooking and drinking varies among individuals, from about 20 ounces to 2 quarts (0.6 to 1.9 liters). The average use is about 3 pints (1.4 liters)—around half as plain water consumed as a beverage and the rest consumed in other beverages (juice, coffee and so forth) and used for cooking.

Because of other community uses (e.g., industrial, commercial, schools, parks and firefighting), your water supplier pumps much more water than is used in households. A recent study of 1,100 U.S. water suppliers showed that to supply all the water needed for all uses, the average amount of water pumped is about 100 gallons (378 liters) each day for each person.

In Canada, the average total home water use is about 230 liters (60 gallons) for each person each day. The use for drinking only has been estimated at about 1.5 liters (1.6 quarts) each day.

83. How long can I store drinking water?

Drinking water that is thoroughly disinfected, such as water from your public water supplier, can be stored for at least six months in capped containers that will not rust or break, such as plastic. Make sure the storage container is completely cleaned before filling. Water that has been boiled for one minute, or three minutes at high altitudes, can be stored for up to one year.

Bottled water should be stored unopened in a cool place; under warm conditions the water may taste like the plastic it is stored in because plastics sometimes leach chemicals. Replace the water every six months and keep it sealed; this will also minimize the "flat" taste that occurs after extended storage. Keep stored water out of the direct sunlight and away from other stored chemicals.

If possible, store water in a refrigerator to help control bacterial growth. Water is not sterile or devoid of living things, but it should be safe from harmful microorganisms. The chlorine residual from your tap water might only last about a month or less in stored water.

84. How much water should I store for emergencies?

A good rule of thumb is to store one gallon of water per person per day. Emergency planning experts recommend storing enough water for at least three days, which means a family of four should store about 12 gallons (45 liters). People with special needs, such as nursing mothers, young children and family members with illnesses, may require more water to be available.

85. How do I treat my water in an emergency?

If the water has been contaminated by living organisms, the most effective method of disinfecting it is boiling. Bring the water to a boil for one full minute (three minutes if you are at a high altitude), then allow it to cool before storing. Use caution because heating and boiling are burn hazards.

When boiling is not practical, you can chemically disinfect your water. Commonly used household chemical disinfectants are chlorine and iodine. Chlorine can be found in common household bleach. Make sure the bleach does not have additives. Check the label: if the available chlorine in the product is around 1 percent, add 10 drops to one quart or liter of water; if it's 4 to 6 percent chlorine, add 2 drops; if it contains 7 to 10 percent, add one drop. If the percentage is not listed, add 10 drops to one quart or liter of water; double that if the water is colored or cloudy. Mix the water thoroughly and allow it to stand for 30 minutes. After this time, the water should have a slight chlorine odor; if not, repeat the chlorine addition and allow the water to stand an additional 15 minutes. If the chlorine taste and odor is too strong, let it stand exposed to the air for several hours or pour it back and forth from one clean container to another to aerate the water and reduce the remaining chlorine. If sediment settles to the bottom, decant off the top layer of water and leave the sediment behind.

Common household iodine from a medicine cabinet or first-aid kit may also be used. Add five drops of 2 percent U.S.P. tincture of iodine to one quart or liter of water. If the water is colored or cloudy, add 10 drops.

Let the water stand at least 30 minutes before use. Iodine tablets are also available from outdoor outfitting stores, as they are sometimes used by backpackers and campers to purify untreated water. These stores also have portable filters.

Note that chlorine and iodine are only somewhat effective in protecting against *Giardia* cysts and may not be effective at all against *Cryptosporidium* oocysts. Therefore, use chlorine or iodine only on groundwater supplies (wells), which are not likely to have these contaminants. Water from rivers, lakes, reservoirs or springs needs to be boiled. A good item to have on hand for emergencies is a ceramic filter, such as those used for camping and backpacking. Remember that boiling or the use of chlorine or iodine do not remove chemical contaminants.

If your water service stops during an emergency, remember the water in your hot water tank, melted ice cubes and the water in your toilet tank reservoir can be used as long as this water was collected before the emergency occurred. If you have the ability to do so, boiling this water is always a good idea before drinking.

86. Should I use hot water from the tap for cooking?

Cold water is best to use. Hot water heaters and plumbing are not designed to preserve water quality. Water in hot water heaters loses its chlorine and promotes the growth of bacteria. Hot water is also more likely to contain rust, copper, and lead from your household plumbing and water heater because these contaminants generally dissolve faster into hot water than into cold water.

On the subject of hot water, insulating your hot water pipes will help the water in them stay warm between uses. So after the first use of the day, hot water will come to the tap sooner, thus conserving water. Hot water should be kept between about 110 °F and 120 °F

(43 °C to 49 °C) to prevent the growth of microorganisms that could be a health issue while reducing the chances of scalding or burns.

87. Should I use hot water from the tap to make baby formula?

No. As noted in the previous question, hot water may contain impurities from the hot water heater and plumbing in your home. Before you draw it, let the cold water run for a couple of minutes if that tap has not been used for a while, overnight or all day, then heat the water in the microwave or on the stove. Catch the wasted water you flush out of the tap in a container and save it for plant watering as a conservation measure. Or better yet, collect fresh drinking or formula water after you have been using a lot of cold water, such as for washing clothes or lawn watering. This will have moved fresh water into your house and you will not have to waste more water to get good tap water.

88. Is it okay to heat water for coffee or tea in a microwave in a foam cup?

Yes. The polystyrene foam in insulated cups is not affected by the microwaves and does not dissolve into the water. As a side note, foam cups shouldn't be called "Styrofoam" cups. According to Dow Chemical, which owns the trademark on Styrofoam, the product is used for "building materials, pipe insulation and floral and craft products, [but] there isn't a coffee cup, cooler or packaging material in the world made from Styrofoam."

89. Is water that comes out of a dehumidifier safe to drink?

No. Dehumidifiers suck water that has not been disinfected from the air. Bacteria and mold spores can accumulate in this water. Also, condenser coils in many dehumidifiers are copper, which could leach into the water. Dehumidifier water is fine for watering plants.

90. How is the water from my refrigerator door different from my tap water?

Your refrigerator should be hooked up to your home's main water supply line, so it is receiving the same water as comes from the tap. However, refrigerator water goes through a point-of-use filter and is chilled in the refrigerator's cooling compartment before being dispensed. As with any POU filter, check with the manufacturer to determine what elements are removed and how often the filter should be changed. If the filter isn't changed regularly, bacteria and other contaminants may build up in the refrigerator cooler.

HOME PLUMBING

91. What should I do to avoid cold-weather problems with my pipes?

First, close all valve connections to the outdoor spigots. Disconnect and drain all outdoor hoses. After detaching the hose, remove any nozzle attachments, and open the faucet to drain the water from the pipes. Insulated caps, available at home improvement stores, can be affixed to the exposed faucets. Use an air compressor to blow remaining water out of underground sprinkling systems only after the valve to your potable water has been closed to avoid air from going back into your house. Close and insulate foundation vents that are near water pipes.

Indoors, insulate pipes or faucets in unheated areas, such as basements or garages, and under your kitchen or bath cabinets as needed. Leave cabinet doors open on extremely cold days to allow the household air to heat the pipes. Make sure you can access your master valve in case pipes do freeze and rupture, and educate everyone in your household on how to shut off the water. If you are expecting severe cold weather and are worried about your pipes freezing, you can also leave a steady, fast drip of water flowing from the tap during the worst of the cold spell. Any extra flow is a severe waste of water.

Also, check with your local water company; you may be responsible for keeping the meter from freezing as well. In other places, meters are maintained by utility personnel.

92. Where do I find my home's master valve?

The most common locations for the master valve in your house or apartment (have your apartment superintendent show you how to do this) are

✓ Where the water supply enters your home
✓ Near your clothes-washer hookup
✓ Near your water heater

To determine if the valve is the correct one, turn it off and see if it shuts off all water faucets in your home. If not, repeat this process with each valve you find until you identify the correct one. If you are unable to locate it or it is inoperable, contact a plumber for assistance. Once you've found the valve, mark it with something distinctive like bright paint, a tag or ribbon so you can locate it quickly in case of an emergency.

93. How can I tell if I have leaks in my home plumbing system?

To check for leaks in your home, first determine whether you're wasting water, then identify the source of the leak. Look at your bill to determine your water usage during a colder month, such as January or February. If your household has four members and your usage exceeds 12,000 gallons (45,400 liters) per month, you probably have some serious leaks. Other ways to determine leakage, according to the USEPA, include:

✓ Checking your water meter before and after a two-hour period when no water is being used. If the meter changes at all, you probably have a leak.
✓ Identifying toilet leaks by placing a few drops of food coloring in the toilet tank. If any color shows up in the bowl before you flush, you have a leak. (Be sure to flush immediately after the test to avoid staining the tank.)
✓ Examining faucet gaskets and pipe fittings for any water on the outside of the pipe to check for surface leaks.
✓ Looking for the triangular arrow on you water meter; if it is spinning, then there is a leak in your plumbing.

You can fix many leaks by replacing worn washers and gaskets, tightening and taping showerheads and replacing toilet flappers.

Some utilities will conduct a free water audit of your home or provide you with a printed survey to help you detect leaks and find ways to conserve water. Contact your utility for more information.

94. How can I tell if my toilet is leaking?

To check, put a few drops of food coloring in the tank, wait about 15 minutes and look in the bowl. If the food coloring shows up there, the tank is leaking and should be fixed. Toilets should be checked for leaks every year.

Some utilities give away conservation kits with food coloring tablets, flow restrictors and plastic bags to fill with water and put in toilet tanks.

95. Why are there aerators on home water faucets?

As the name suggests, aerators add air to water as it flows out of the tap. When mixed with water, the tiny air bubbles prevent the water from splashing too much. The added air also enhances the taste of the drinking water, making it less flat. And because water flow is diminished through an aerator—often reducing it to half the regular flow—aerators also help conserve water.

96. Why do hot water heaters fail?

All waters have natural corrosive properties, so water heaters will eventually rust from the inside to the outside. The time it takes for this to happen varies depending on how corrosive your water is, what has been done to mitigate the corrosiveness and the original quality of the water heater. Because corrosion is also a factor in water pipes, many water suppliers take preventive steps to decrease the corrosiveness of the water they deliver.

In hard water areas, the minerals causing hard water tend to form a scale at the bottom of the hot water heater, which can result in failure of the unit. Using a water softener should help alleviate this problem. Occasionally, flushing the water heater from the bottom will prevent some, but not all, of the scale from forming. Using a water softener should minimize this buildup. Also, iron and manganese can accumulate in the bottom of the heater. Nevertheless, you should drain the water heater periodically to remove scale, rust and mineral particles. However, turn off the heating element for a while before draining to allow the water to cool; if this is not possible, use care because the water is hot or have a plumber do this for you.

97. What causes the banging or popping noise that some water heaters, radiators and pipes make?

Each noise has a different cause. In a water heater, some nontoxic minerals in the water form a rough coating on the inside of the heater when the water warms up. When the container walls are rough, air bubbles form as the water heats. When these bubbles burst, they cause a popping noise. A new water heater will have smooth walls, without scale buildup, so the water will not form bubbles as it heats, and no noises will be heard. In a home radiator heated with steam, the banging noise is caused by the condensed water (steam) pooling in the bottom of radiator. Open the inlet valve all the way so the water can run out to eliminate this.

Pipes make noises for two reasons. First, when you open a hot water tap after water hasn't been used for a while, the pipe leading to 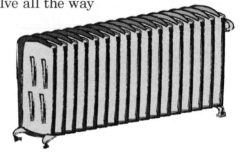 the tap will be cold. As the hot water runs through the pipe, the pipe heats up and expands. This will sometimes cause the pipe to creak or make similar noises.

The other cause for noise in the pipes is water hammer. When water—hot or cold—is moving fast through a pipe and the flow is stopped quickly, the water keeps moving for a while, like a train plowing forward during a wreck. The moving water finally bangs against (hammers) a faucet or valve, making a loud noise, like a hammer hitting metal. If you've noticed this problem in your home, it can easily be corrected by turning the water off more slowly, by installing small standpipes (purchased at plumbing supply stores) on the affected pipes or by padding pipes at appropriate locations—where they contact another object—to reduce loose pipe rattling.

98. There is a blue-green stain where my water drips into my sink. What causes this?

Blue-green stains come from copper, which probably leached from your home plumbing and dissolved in the tap water. The conditions that cause copper in the water also can introduce lead into drinking water, and high amounts of either lead or copper can cause health problems. Call your local water supplier to discuss what actions can be taken to reduce the level of copper and lead in your water. If your water is from a private well, have your tap water tested for lead and copper. To clean the sink, check with your local hardware store for stain-removal products.

99. Why does my water sometimes have sand in it?

Routine cleaning of pipes that carry drinking water can stir up material that has settled to the bottom of the pipes. This can give your water a temporary sandy appearance. Some wells have specially designed screens (like window screens) to hold back the sand, and a break in this screen is another possible cause of sandy water. Small quantities of sand may pass through a well screen, even if it is not broken. The best way to solve this problem is to verify with your water supplier that there is no break in the system, and if there is none, flush your home pipes by running water for a while through your largest faucet, probably in the bathtub.

Saving this water for other uses is a good conservation measure. Some particles that look like sand may actually result from the corrosion of galvanized pipes in your household plumbing.

HOME TREATMENT

100. Should I install home water treatment devices?

If you draw your water from a private well, you may want to consider installing a point-of-use (POU) or point-of-entry (POE) home water treatment device. However, water delivered by a public water system, be it a private company or a municipal supplier, must meet stringent requirements for chemical, physical and biological contaminants that could cause illness, while home treatment devices generally do not need to meet federal, state or provincial drinking water standards.

In some small communities, installing sophisticated treatment technologies to meet the regulations may be too expensive for the ratepayers to bear. In these cases, in-home water treatment systems may provide a low-cost alternative. USEPA has authorized the use of home treatment equipment to meet requirements of the Safe Drinking Water Act regulations for some contaminants. If this is the case in your community, your water utility may provide you with a POU or POE unit or information on how to obtain, install and maintain a POU or POE device.

Otherwise, use of POU or POE equipment is a personal decision. If you are concerned about aesthetic qualities such as color, taste, odor or hardness or desire additional barriers against disinfection by-products or residual traces of contaminants such as lead or *Cryptosporidium*, you might consider a home treatment unit. Be sure that you read the information about the product carefully, or talk to a qualified distributor, so you get the right product for the concern that you have, and check to see if the device is certified.

If you do decide to use a home treatment device, you must be careful to maintain it properly according to the manufacturer's instructions. For example, modern POU and POE treatment products sometimes shut off automatically when a filter is depleted or warn users when replaceable cartridges need changing or routine maintenance is needed.

101. How do I know which type of POU/POE to use?

POE systems are installed at the service line where the water comes into the house, whereas POU systems are placed in several locations within the home: countertop, faucet-mounted, under-sink cold tap or under-sink line bypass. Treatment units can be grouped into six general categories.

Particulate filters can remove waterborne particles, including black manganese particles that cause staining, of different sizes.

Adsorption units usually contain activated carbon (sometimes incorrectly called activated charcoal or just charcoal) and may remove chlorine, tastes and odors and organic compounds such as disinfection byproducts. Microbes can grow in these units, but by following the manufacturers' guidelines, microbes can be controlled. Adsorption units are generally not designed to remove copper and lead. Certain special filters will remove dissolved lead, but check their claims with independent organizations.

Oxidation/filtration systems change iron (clear water turning red) or hydrogen sulfide (the rotten-egg odor) into a form where these nontoxic but troublesome chemicals can be filtered out of the water. These are primarily POE systems and are frequently used by people who have their own source of water, such as a private well.

Water-softening systems soften water by trading (exchanging) nontoxic calcium and magnesium that cause hard water for other nontoxic chemicals that do not cause hardness, such as sodium. These units must be renewed (regenerated) periodically with salt.

Reverse osmosis units remove hardness; chemicals such as nitrates, sodium, dissolved metals (such as lead and copper) and other minerals; and some organic chemicals. Reverse osmosis units also remove fluoride. Some units are sensitive to chlorine, so a chlorine-removal step usually is included prior to the reverse osmosis unit. Reverse osmosis units do allow some organic chemicals to pass into the treated water, however. Therefore, sometimes these systems are followed by adsorption units to remove these organic compounds. Reverse osmosis units are usually POU devices that treat relatively small volumes of water. Remember that if the equipment removes the disinfectant in your tap water, the water must be stored properly to avoid subsequent contamination.

Distillation units boil the water and condense the steam to create distilled water, removing some organic and inorganic chemicals (such as hardness, nitrates, chlorine, sodium and dissolved metals). Distillation units also remove fluoride. However, some organic chemicals may pass through the units with the steam and contaminate the distilled water unless the unit is specifically designed to avoid this problem.

All POU and POE units require maintenance and should be bought from a reputable dealer. Their performance should be tested and validated against accepted standards like those used by NSF International, the Water Quality Association and Underwriters Laboratories. These standards allow manufacturers' performance claims for drinking water treatment unit products to be tested and checked. You can investigate claims made for any unit by visiting the Internet site of the testing organization named in the product's claims or by contacting the organization directly.

✓ NSF International: www.nsf.org/consumer/water, (630) 505-0160
✓ Water Quality Association: www.wqa.org, (800) 749-0234
✓ Underwriters Laboratories: www.ul.com/water, (847) 272-8800

USEPA has an environmental technology verification program that has tested numerous products for their effectiveness in removing certain contaminants. For reports on individual products, go to www.epa.gov/ nrmrl/std/etv/vt-dws.html.

102. Do water treatment devices using electromagnets work?

While several companies selling magnetic treatment devices, particularly for water softening, have advertised on the Internet in recent years, their claims have not been verified by reliable scientific study. According to a Magnetics Task Force Report issued by the Water Quality Association, a nonprofit organization representing home, commercial and light-industrial water treatment industries, there is no scientific evidence to determine the effectiveness of physical water treatment technologies such as magnetic, electromagnetic and catalytic devices in home water treatment devices. A reporter for the *Skeptical Inquirer* magazine reviewed more than 100 scientific studies of the technology and concluded, "Much of the available laboratory test data imply that magnetic water treatment devices are largely ineffective."[1]

---

1 Powell, Mike. 1998. Magnetic Water and Fuel Treatment: Myth, Magic, or Mainstream Science? *Skeptical Inquirer.* 22:1

103. Is distilled water the "perfect" drinking water?

No. Distillation involves boiling the water, so distilling removes harmful bacteria and viruses, *Giardia* cysts and *Cryptosporidium* oocysts, and many nuisance minerals, as well as harmful chemicals like lead, copper, nitrates, sodium, some organic contaminants and chlorine. But distillation also removes the water's natural minerals, leaving the water "flat" tasting. And, some organic contaminants like chloroform and cleaning fluid (solvents) may leave the water with the steam and end up in the final water when the steam is cooled, so most water distillers 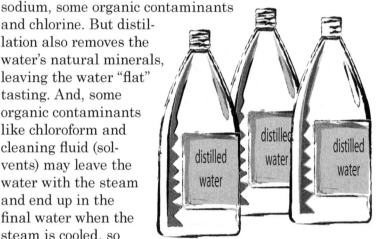 have added treatment to prevent any organics in the steam from ending up in the final product.

Many people keep a store-bought container of distilled water for use in steam irons and car batteries and for watering plants. And because most of the minerals are missing, using distilled water to make tea or coffee will avoid a buildup of white calcium–magnesium scale on the kettle or pot. A home water softener will also take care of this problem.

Except in special cases, such as removing salt from seawater to make drinking water, distilled water is too expensive for your public water supplier to treat large volumes of water, and it is really not necessary to treat your water to such an extent.

HARD WATER AND SOFTENING

104. What is "hard" water?

Hardness in drinking water is caused by calcium and magnesium—two nontoxic, naturally occurring minerals in water. If calcium or magnesium is present in your water in substantial amounts, the water is said to be hard because lathering soap for washing is difficult to do, and cleaning with hard water is inefficient. Water containing little calcium or magnesium is called soft water and is better for laundering and other purposes.

Other indications of water that is excessively hard include clothes that look dingy and feel harsh and scratchy after laundering; spots on dishes and glasses after they've been washed; filmy shower doors; and sticky, dull hair, even after washing. A buildup of mineral deposits can also reduce water flow in domestic pipes, and faucet aerators may become plugged if not cleaned often.

105. Should I install a water softener in my home?

If very hard water is interfering with your laundry or personal and home grooming, a water softener can help. You can find out the degree of hardness of your drinking water by contacting your water supplier or by testing it yourself with a home water test kit. The hardness of your water will be reported in grains per gallon, milligrams per liter (mg/L) or parts per million (ppm); one grain of hardness equals 17.1 mg/L or ppm of hardness. The higher the hardness number, the more a water softener will help.

USEPA Hardness Scale		
Classification	**mg/L or ppm**	**grains/gal**
Soft	0 – 17.1	0 – 1
Slightly hard	17.1 – 60	1 – 3.5
Moderately hard	60 – 120	3.5 – 7.0
Hard	120 – 180	7.0 – 10.5
Very Hard	180 & over	10.5 & over

A water softener can reduce the formation of scale in your hot water system and make washing easier. Water softeners replace the calcium and magnesium with sodium or potassium, which dissolve in water and do not leave deposits.

Water softeners are regenerated with salt. After the salt is used, it goes down the drain and into the environment, by the way of a wastewater treatment plant. It's best to regenerate a softener after a set amount of water has passed through it rather than on a particular time schedule to avoid wasting salt by regenerating it too soon or using the softener after it no longer is effective. Manufacturers' directions should provide this information.

You may consider installing the softener only on your hot water line to save money and benefit the environment. Softening only hot water means less water passes through the softener, which means it needs regeneration less often and less salt is used and recycled into the waste stream. And, if cold water isn't softened, a cold glass of tap water still has the taste and health benefits of the water's mineral content. Softened water used for outside irrigation is also not a prudent use.

Softening only the hot water does have a couple of disadvantages. First, if you wash your clothes in cold water, you won't get the benefit of soft water; but you can buy softening products to add to your wash. Second and more importantly, if your water is very hard, the water will still be fairly hard when you mix hot and cold water together, and you will see only minimal benefit from the softener.

106. Does softened drinking water have any negative health effects?

The amount of sodium or potassium in the water after softening is relatively insignificant when compared to the amount found in many foods, so it should not be a problem unless you are on a very sodium-restricted diet. And contrary to some urban legends, drinking softened water does not drain calcium and magnesium from the body.

107. I have a water softener, but I still get spots on my bathroom tile and dishware. Why?

All water contains dissolved nontoxic minerals. Calcium, magnesium, sodium, sulfate, chloride and bicarbonate are the most common. A water softener exchanges calcium and magnesium for sodium or potassium, so the water leaving the softener contains no calcium and magnesium (thus, no hardness) but more sodium or potassium.

Therefore, all minerals are not eliminated during softening, just traded for other minerals. If you put the softened water in a dish to completely evaporate, the white stuff left over—although it would be different—would look the same and would equal the same amount as before the softener was installed.

108. How can I get rid of the precipitate deposits on my coffee or tea pot and showerhead?

To remove a buildup of calcium and magnesium and other deposited minerals, fill your coffee pot with vinegar and let it sit overnight. Soak the showerhead overnight in a plastic bowl filled with vinegar. When you are done, carefully discard the contents of the plastic bowl down a drain, and flush the container and sink drain with plenty of water. Rinse the coffee pot or showerhead thoroughly after treatment and before use. After the deposits are gone, help keep the buildup to a minimum by pouring the excess hot liquid out of your coffee pot when you are finished using it, and wiping off the showerhead with a soft cloth.

White spots on glass shower doors are difficult to remove with vinegar because the spots dissolve very slowly. Commercial products are available, but handle these carefully and follow the manufacturer's directions. To prevent spots from forming, wipe the glass door with a damp sponge or towel after each shower. Home water that has been softened will likely produce fewer spots as well as spots that are easier to remove.

109. Some of my clear glassware comes out of the dishwasher with a rainbow sheen on it. What causes this?

Too much dishwasher detergent combined with soft water can leave an oily looking sheen on glassware. Also, some detergents can actually etch glasses with high silica content, resulting in a fine pattern of etching that appears as a rainbow coating.

BOTTLED WATER

110. Should I buy bottled water?

Many people turn to bottled water as a convenience, and in the vast majority of cases, both bottled water and tap water are safe, healthy choices. But if your tap water meets all of the federal, state or provincial drinking water standards, you don't need to buy bottled water for health reasons. Or, if you want water with a different taste or without health additives such as chlorine and fluoride, you can buy bottled water. However, there has been some concern that children will not get healthy levels of fluoride by drinking bottled water. You should check with your dentist.

Bottled water is popular; Americans spend about $6 billion annually on this product, and it is now second only to soft drinks in popularity. Recent surveys show that almost 70 percent of people in California, for example, drink bottled water at least some of the time. Of course, in emergencies, bottled water or stored tap water can be a vital source of drinking water for people without water.

Cost and environmental protection are two major reasons to carefully evaluate choosing bottled water. Bottled water can cost up to 1,000 times more than municipal drinking water, which is about 0.004 cents per gallon ($0.001 per liter) compared as much as $8 per gallon ($2.11 per liter) for bottled water. On the environmental side, 1.5 million tons (1.36 million metric tons) of plastic are used each year to bottle water, and much of that ends up in landfills. Along with the energy used to produce and recycle the plastic, energy is used to transport the product to stores around the world. While the International Bottled Water Association says that bottlers are protective of water resources, some manufacturers may be pumping up to 500 gallons per hour (1,900 liters) from valuable aquifers and other water sources.

Remember, if you use bottled water, consider it a food, refrigerate it after opening and store it away from direct sunlight and other chemicals.

111. Is bottled water regulated?

In both the United States and Canada, bottled water is regulated as a food product, which means it is regulated at the production source, but not once the product has gone into distribution channels. The U.S. Food and Drug Administration (FDA) is the U.S. regulator, and the Canadian Food Inspection Agency (CFIA) monitors bottlers under the Food and Drugs Act while Health Canada establishes health and safety standards for bottled water and packaged ice and develops labelling policies related to health and nutrition in that country. Provinces and states may regulate the bottled water sources, including drilling and construction practices, allowable rates of production and watershed protection.

The FDA has established limits for more than 75 microbiological, physical, chemical and radiological substances for both the source water and the finished bottled water product, and these standards must be as stringent as USEPA standards for public water supplies. Bottlers are subject to annual inspections and must test weekly for the presence of bacteria in their water. Individual states and industry trade groups may impose additional regulations and requirements on bottlers. Regulations also require manufacturers' labels to list the contents and source of the bottled water.

In Canada, bottled water companies must adhere to similar quality standards, good manufacturing practices and labeling requirements.

112. What do the labels on bottled water mean?

The label identifies the source of the water, such as spring, artesian well or municipal water supply, and the type of water, such as mineral, purified or sparkling.

The FDA also requires nutritional information to be listed on the label of any food product, which includes total calories, calories from fat, sugars, protein and fiber. Except for designer water or enhanced "sports" or "vitamin" waters, these elements are not found in most waters. If the water contains more than an acceptable amount of a regulated substance, this must be stated on the label (e.g., "This product contains excessive amounts of iron"). Particular types and sources of water are described below.

Artesian water or artesian well water is water that comes from a well drilled into a confined aquifer in which the water flows freely to the top of the well under natural water pressure.

Groundwater is water from an underground saturated zone—an aquifer—that is under a pressure equal to or greater than atmospheric pressure. Groundwater used for drinking water that is under the direct influence of surface water must be treated.

Well water is water from a hole that is bored, drilled or otherwise constructed in the ground that taps the water of an aquifer.

Spring water is collected from an underground formation from which water flows naturally to the surface of the earth. Water must be collected at the spring or through a borehole tapping the underground formation feeding the spring. A natural force must cause the water to flow to the surface, and the location of the spring must be identified.

Mineral water is water that contains 250 milligrams per liter (mg/L) or more of total dissolved solids, determined by evaporating the water and weighing the residue. Mineral water comes from a source tapped at one or more boreholes or springs, originating from a geologically and physically protected underground water source. Some mineral waters are naturally carbonated.

In Canada, bottled water classified as mineral or spring water cannot have its composition modified through the use of any chemicals except for the addition of fluoride, carbon dioxide or ozone.

Purified water or demineralized water is water from any source that is processed by distillation, deionization, reverse osmosis or other processes to remove impurities. Alternatively, water may be called deionized water if it has been processed by deionization; distilled water if it has been processed by distillation; reverse osmosis water if it has been processed by reverse osmosis, and so forth. Sometimes water treated to this level must be remineralized to regain a pleasant taste.

Sparkling water is water that is enhanced with gaseous carbon dioxide—it's carbonated and contains small bubbles and an effervescent texture. Seltzer water and club soda are also carbonated, but they are considered soft drinks.

Sterile water or sterilized water is water that meets the requirements of the sterility tests in the United States Pharmacopeia. The FDA requires the following additional labeling requirements:

✓ If the total dissolved solids content of mineral water is below 500 mg/L or above 1,500 mg/L, the statement "low mineral content" or "high mineral content" must be added to the label, respectively.
✓ If the source of the bottled water is a community water supply or a municipal source, the source must be on the label, unless the water has gone through further treatment to render it as "purified" water.
✓ If the product states or implies that it is to be used for feeding infants and the product is not commercially sterile, "not sterile" must be added to the label.

✓ If the product does not meet the standards discussed above, the terms "contains excessive bacteria," "excessively turbid," "abnormal color," "abnormal odor," "contains excessive chemical substances" or "contains excessive radioactivity" must be applied to the label as appropriate.

Check the label carefully on any bottle of water to determine the sodium content, regardless of the general labeling. Some bottles labeled "sodium-free" do contain some sodium, which may be of concern for those on a sodium-restricted diet. To be safe, choose a brand of bottled water that contains zero milligrams (mg) of sodium per 8 ounces (0.24 liters).

113. Is bottled water okay to store?

Yes, to a limit. Bottled water is a good source of drinking water during emergencies and when a person is on the go, but it does not store well because bottlers often remove or do not add any residual chlorine to the water, which would provide some protection against microbes over time. Unless stated otherwise, bottled water is not sterile. Ozone is the preferred disinfectant treatment used by bottled water manufacturers, but reverse osmosis—ultraviolet light, distillation, activated carbon, cation exchange and microfiltration are also used. Accordingly, a chlorine disinfectant is not used, so microbes can grow in the water over time. These microbes are not germs, so they will not make you sick, but to maintain freshness, the International Bottled Water Association recommends that water bottles be labeled with the bottling date and be replaced every six months. Bottled water has a designated shelf life in a store, as well.

DOWN THE DRAIN

114. Where does the water go when it goes down the drain?

If you are on a public sewer system, all of the drains in your house are connected to a single pipe that leads to a main collection pipe under the street where the wastewater is collected from all the homes in your area and transported to a larger pipe that collects water from other streets. The wastewater then flows into still bigger pipes that connect various neighborhoods.

The pipes in the wastewater collection system are larger and contain more liquid the nearer they are to the wastewater treatment plant. Here, the wastewater is treated and cleaned so that it can be put back into the environment without harming anything.

If you are not connected to a sewer system, the liquid wastes from your home go into a septic tank, where most of the solids settle out. The water then overflows to a leach field where the pipes with holes in the bottom are buried in the ground. The water seeps out of these holes and into the ground where it is naturally recycled.

115. What household chemicals can I safely pour down the sink or into the toilet?

Before you think about what you can throw away, think about what you are buying. Start by buying environmentally friendly products whenever possible. Next, try to buy just what you need so you won't have any or much left to dispose of. Finally, check with your local solid waste department or similar agency to learn of local rules and hazardous waste collection days. In Canada, the local sewer-use ordinances control disposal in most municipalities.

If your home is on a public wastewater system, these liquids can safely be poured down a drain at a time when you have been using a lot of water (such as washing clothes) to help dilute the wastes:

✓ Aluminum cleaners
✓ Ammonia-based cleaners
✓ Drain cleaners
✓ Window cleaners
✓ Alcohol-based lotions
✓ Bathroom cleaners
✓ Depilatories
✓ Hair relaxers
✓ Permanent lotions
✓ Toilet bowl cleaners
✓ Tub and tile cleaners
✓ Water-based glues
✓ Paintbrush cleaners with trisodium phosphate
✓ Lye-based paint strippers

After disposal, be sure to rinse the empty container with water several times. Of course, the safest course of action is not to put anything in your sink or toilet. Other wastes should be disposed of by following your local municipality's hazardous waste collection process.

116. How do I safely dispose of pharmaceuticals?

Don't flush pharmaceuticals down the toilet or pour them down the drain because wastewater treatment plants and septic systems are not designed to treat pharmaceutical waste and the drugs often end up in our waterways. The US government has recently established alternative guidelines for disposing of prescription medicines:

✓ Take unused, unneeded or expired prescription
drugs out of their original containers and throw
them in the trash.

✓ Mix prescription drugs with an undesirable substance, such as used coffee grounds or kitty litter,
and put them in impermeable, nondescript containers, such as empty cans or sealable bags, to further
ensure the drugs are not diverted.

✓ Flush prescription drugs down the toilet only if the
label or accompanying patient information specifically
instructs doing so.

✓ Take advantage of community pharmaceutical take-
back programs that allow the public to bring unused
drugs to a central location for proper disposal.

✓ Take medications with attached needles to hospitals
or other approved sites for proper disposal.

In Canada, the only allowed and encouraged option
is to return unused pharmaceuticals to the pharmacy.
Most provinces have programs for this.

117. I have a septic tank. Should I take any special precautions?

If you have a septic tank, check with your local authorities for any regulations or special conditions you should
be aware of, but here are some general rules. First,
remember that any substance put down the drain into a
septic system may eventually seep into the local groundwater. Do not use any chemicals to clean your system;
they may actually harm the system or the groundwater.
Second, don't bother with septic-tank additives or the
addition of yeast because these elements really don't
help the septic tank very much.

Third, minimize water usage. Don't run water con-
tinuously while rinsing dishes or thawing frozen food
products (these are good conservation measures in any

household). Consider limiting toilet flushes or putting
a plastic bottle full of water in the toilet tank to reduce
the amount of water used in each flush. Run only full
loads when using a dishwasher or washing machine. Try
to use the washing machine at times when water is not
being used for other purposes. The reason for all of this
is to pace the flow of waste through your septic system,
allowing it time to do its job effectively.

Fourth, do not dispose of these items down the drain
or toilet:

✓ Fats, grease or cooking oil
✓ Coffee grounds, meat bones or other food products
  that don't biodegrade easily, even if they have gone
  through the garbage disposal
✓ Household cleaning fluids
✓ Automotive fluids such as gas, oil, transmission or
  brake fluid, grease or antifreeze
✓ Pesticides, herbicides or other potentially toxic
  substances
✓ Nonbiodegradable substances or objects such as
  cigarette butts, disposable diapers and feminine
  hygiene products

Finally, do not connect any footing or foundation
sump pumps to the septic tank system. In all cases, follow your municipal or township codes and regulations.

FISH AND PLANT LIFE

118. How should I fill my fish aquarium?

First, allow at least 1 gallon (4 liters) of cold water
to run from the tap before using the water to fill the
aquarium to flush any copper or zinc from copper or
galvanized piping in your home (or collect the water

after a lot of water has been used for other reasons such as showering or washing clothes). Tropical fish are very sensitive to small amounts of copper or zinc in their water. Then, hold a plate above the aquarium and pour the water onto the plate from about 1 foot (30 centimeters) above it before it hits the plate and flows into the tank. This adds air (oxygen) to the water. Let the water sit in the aquarium for an hour or two until it reaches room temperature.

If your water comes from a municipal supply that is disinfected with chlorine or chloramines, consult your local pet store to learn how to test for and remove the disinfectant in the water. Chloramines in particular are toxic to marine organisms. Remove the disinfectant from the water in the aquarium before adding the fish. If chlorine cannot be removed, it will dissipate if the water is allowed to stand for a few days before adding fish. For tropical fish kept in a seawater environment, maintain the appropriate salt level using a saltwater additive and check concentration using a specific gravity indicator.

119. I have trouble keeping fish alive in my fish pond. Is there anything I can do?

Domesticated fish die for many reasons. One problem is that waste products from the fish can decay and release ammonia, which is quite toxic to the fish and other aquatic life. Commercial products are available that will capture the ammonia, and other harmful chemicals such as nitrate and nitrite, while maintaining the appropriate pH level. If you use a biological filter, the ammonia will be changed to a nontoxic chemical by microbial action. If you fill the pond with tap water, you may want to follow procedures similar to those used in filling an aquarium. Disease is also a possibility. Check with your local pet store for more advice.

If your fish are disappearing, it could be that raccoons or fishing birds, such as herons, have stolen them.

120. When I try to root a plant or grow flowers from a bulb in my house, the water looks terrible after awhile. What will prevent that?

Put 1 to 2 tablespoons of activated carbon (sometimes called agricultural charcoal) in the bottom of the bowl. This will help keep the water in better condition. Of course, changing the water frequently also helps.

121. Roses, azaleas, camellias and rhododendrons all require acid conditions. How should I adjust the acid content of my plant water?

The acid content of water is measured by the pH. Any number below 7.0 indicates that the water is acidic, and pH levels above 7.0 are alkaline. Some plants like water with a pH somewhere between 6.5 and 6.8, lower than found in most tap waters. Acidic fertilizers, vinegar or a drop of hydrochloric acid (handle with care) can be used to lower the pH of tap water before pouring it on these sensitive plants. Before deciding how much acid to add to a gallon of water, do some testing using a pH test kit that can be obtained from garden or pool supply stores. Also, check with the water supplier as to the pH and alkalinity of the tap water. The higher the alkalinity the more difficult it will be to change the pH.

122. I live in a very hard water area and I have a water softener. My plants don't seem to like my tap water. What can I do?

Water softeners usually replace the hardness chemicals (calcium and magnesium) with sodium. If you soften very hard water, you will wind up with a fair amount of sodium in your tap water, and some plants don't like sodium. Discuss this with your local garden store, or try one of these alternatives:

✓ Use reverse osmosis-treated water or distilled water for watering plants.
✓ Change your softener regeneration chemical from sodium chloride to potassium chloride (although this may be more expensive and harder to get).
✓ Collect water for the plants from a tap before the water gets to the softener, for example, from an outside garden hose connection.
✓ If you are using sodium-softened water on plants, water heavily to rinse off previously deposited minerals and discard excess water. Heavy sodium salt concentrations in the absence of calcium and magnesium may cause the dirt to swell a bit and retard the growth of plant roots.

123. What causes the whitish layer on the soil of my potted plants?

Drinking water can contain many nontoxic chemicals and minerals. When the water on your plants evaporates, these chemicals—no longer dissolved—are left behind as a whitish layer. Using distilled water on your plants will avoid this problem. You can also catch rainwater for watering plants, but cover any water that is kept outside to prevent insect larvae infestation. If you have a dehumidifier, the water that comes out of it is good for watering plants. Distilled water, rainwater and dehumidifier water will wash minerals out of the soil, however, so use a slow-release balanced fertilizer to replace these essential elements.

COSTS

124. What is the cost of the water I use in my home?

Most people pay for water delivered to their home according to the amount they use. In the United States, the water rate is usually based on each 1,000 gallons used; in most other industrialized countries, the charge is for each cubic meter (m³) used. This is the "volume" charged for actual water used. One thousand gallons (3.8 m³) of water typically serves one consumer for about 20 days. Prices vary greatly, but a typical cost in the United States is about $4 for 1,000 gallons (3.8 m³). In Canada, where households tend to use less water, typical rates are about $1.12/m³ (264 gallons) in Canadian dollars, so rates are about the same in both countries.

Most utilities add a $5 to $10 per month base charge for fixed utility costs in the distribution system (the cost to maintain the infrastructure to get the water to your home). In some places, this is called a *meter charge*. Still, either a gallon or a liter of tap water costs less than one penny.

Of the amount charged for 1,000 gallons, about $0.45 to $0.75 is for treatment; the rest is for paying the capital costs of the treatment plant and the pipes in the street, as well as the salaries of the employees who work for the drinking water utility and other fixed costs.

You can calculate the cost of water in your area by looking at your water bill and dividing the total cost for water by the total amount of water used (just use the water part of the bill if other costs are included). In general, in the United States we spend about 0.5 percent of our income on both drinking water supply and waste-water disposal.

125. How does the water utility know how much water I use in my home?

Most buildings are equipped with a water meter that measures the amount of water used. Some community water systems—particularly in smaller communities—are not metered, so households are charged the same flat rate each month. Financial experts consider this a less equitable and less conservative rate system. In communities with water meters, a water utility employee reads the meter on a regular schedule, either in person or electronically from a remote location or passing vehicle. The previous reading is subtracted from the current reading to determine the amount of water actually used. And in a few communities, the cost of water is covered by general taxes.

126. How does the water company know that my water meter is correct?

Most water companies routinely test water meters on a rotating basis to ensure that the readings are accurate. If you notice that your recorded water use changes suddenly for no obvious reason, such as from having more people in your home, an extended trip away from home, or heavy lawn watering, report this to your water supplier so it can be investigated. Sometimes an undetected leak in the home can increase your water bill unexpectedly. In most instances, when a water meter is wrong, it reads low. As a good citizen, you should report this to your water supplier just as you would when you think your meter might be reading too high.

You should aso know that meters wear out in time, and as they do, their accuracy diminishes. So if your meter is replaced, don't be surprised if your bill goes up slightly as the meter reading is more accurate.

127. We had a conservation drive in our area and everyone cooperated. Then our water rates went up. Why?

Water suppliers have fixed costs—salaries, hydrant maintenance, mortgages and so forth. They must collect this money regardless of water use, so when water volume goes down because of conservation by the public, the cost of each gallon of water used sometimes is raised to provide the water supplier with the money it needs to maintain its system. However, water conservation can eventually lead to stable rates for a longer period of time, because capital improvements or new investments can be postponed if demand does not increase.

Sources

We all live down stream,
*   better watch out what we put in it*
We all live downstream,
*   let's be careful what we do*
We all live downstream,
*   better look out how we're using it*
We all live downstream,
*   and the stream is running through*
*   — Banana Slug String Band,*
*   "We All Live Downstream," 2008*

128. Where does my drinking water come from?

Drinking water comes from two major sources: water that flows aboveground, known as *surface water*, and water that is pumped from beneath the ground, called *groundwater*. Surface water comes from lakes, reservoirs and rivers. Groundwater comes from wells that water suppliers drill into aquifers—underground geologic formations of permeable rock. Wells less than 100 feet (30 meters) deep are considered shallow, while deep wells can extend 1,500 to 2,000 feet (450 to 600 meters) below ground.

Springs are another source of freshwater. Springs begin underground as groundwater. When the water is pushed to the surface and flows out of the ground naturally, it becomes a spring. The water then may flow over the surface of the ground as a creek or river, or may form a lake.

The majority of people in the United States—66 percent—live in areas served by large water systems that rely on surface water. The larger Canadian cities, including Montreal, Toronto, Edmonton and Vancouver, also use surface water.

However, 80 percent of public water systems are in smaller communities that rely on groundwater. About 80 million people in the United States use groundwater supplied through individual wells and municipal groundwater systems. Another 2 million Canadians are supplied by municipal groundwater systems.

If you get your water from a public system, your local water utility explains the specific source of your drinking water in its annual water quality report to customers, and this information often is available on a utility's Web site.

129. How does nature recycle water?

Through the processes of evaporation, condensation, precipitation and infiltration—the hydrologic cycle—the total amount of water on the globe remains constant. Water from oceans, lakes, rivers, ponds, puddles and other water surfaces evaporates to become clouds. The clouds make rain, snow or sleet that falls to earth to create rivers and streams, or seeps into the ground to form groundwater. All of this water flows to the ocean to start the cycle over again.

130. I have heard the term "mining groundwater." Is that anything like mining coal?

Yes, mining water is similar to mining coal. Just as when coal is mined, less and less remains in the ground; the same is true of water after water mining. Although groundwater is replenished by rainfall soaking into the ground, the process is slow. When groundwater is pumped out of the ground faster than it can be restored, the groundwater is being mined. When this happens, the underground level of the groundwater falls, and wells have to be drilled deeper to reach it.

131. Can groundwater be restored in any other way than the natural water cycle?

Yes. In recent years, some water suppliers have adopted a system called *groundwater recharge*. This is often done in large basins where volumes of water spread over the land seep back into the aquifer. The water can be ultra-treated wastewater, diverted river water, surface runoff or other excess water. This reused or excess water can also be returned to the aquifer through direct injection into recharge wells. On Long Island, New York, some groundwater used for air conditioning is returned to aquifers through recharge wells. In Arizona, unused Colorado River water that the state owns is delivered through a series of aqueducts and is recharged into the ground for later use. This is called *water banking*.

Some water utilities have also adopted aquifer storage and recovery programs, which use similar recharge methods to inject treated drinking water into aquifers.

132. How much drinking water is produced in the United States each day, and how does that compare with the water produced and used for industrial purposes and irrigation of crops?

Based on the weight of product manufactured, the water supply industry is by far the largest manufacturer of "product" in the United States and Canada. In just 16 hours, U.S. water utilities produce as much water in tonnage as the oil industry outputs in a year; in 24 hours, as much water tonnage is produced as the steel industry produces in an entire year; and in a week, the tonnage of water produced is equal to the yearly output of all of the bituminous coal producers—truly a monumental daily quantity of product.

This amounts to 40 billion gallons (227 billion liters) of tap water daily for domestic use (homes, restaurants, hotels, small businesses and so forth) in the United States, 60 percent from surface water and 40 percent from groundwater.

Daily irrigation use is much larger, depending on the location and the time of year. It takes about 50 glasses of water just to grow enough oranges to produce one glass of orange juice. One estimate puts the total amount used for irrigation at 141 billion gallons (554 billion liters) a day, 66 percent from surface water and 34 percent from groundwater. Of course, crop irrigation water is diverted from the source and does not get treated as tap water is.

The average American couple uses 200 gallons (760 liters) of water in their home each day, which translates to more than 1,660 pounds (753 kilograms) of manufactured product each day, about 300 tons (272 metric tons) each year.

A Canadian household of two uses about 165 metric tons (180 tons) of water each year.

133. How much of the Earth is covered with water, and how much of that is drinkable?

Close to three-quarters of Earth's surface is covered with water, but less than 1 percent is suitable and available for drinking water using conventional water treatment.

Of all the water in the world, 97 percent is in the oceans and about 1.7 percent is locked up in ice and snow; ice caps and glaciers hold the majority of total freshwater on Earth, about 68.7 percent. Another 0.3 percent of fresh water is in aquifers that are too deep to access, and about half of the 0.3 percent of groundwater that is accessible is too salty to use efficiently.

134. What country has the most potable water?

If money were made out of water, Finland would be the richest country in the world, according to the Water Poverty Index compiled by Keele University in the United Kingdom. The ranking measures water availability in a given country and people's capacity to access that water, and includes the reliability or variability of the resource. Because of its plentiful supply of surface water and groundwater, an infrastructure that provides ready access and several other factors, Finland ranked richest among 147 countries studied, and Canada was second. Haiti was last. Penalized for what a panel of water experts saw as its wasteful use of water, the United States ranked 32nd.

135. Can ocean water be treated to make drinking water?

Ocean water contains so much salt that at least 99.2 percent of the salt must be removed to avoid a salty taste in drinking water, but it can be treated to make drinking water through a process called *desalination*. Many people around the world, particularly in arid countries in the Middle East and Africa, as well as in coastal communities in North America, rely on this process. However, the process is more expensive than traditional treatment because it requires a lot of energy. The cost of desalination has been estimated at $5 to $10 for each 1,000 gallons ($0.85–$1.90 per 1,000 liters) instead of the average of $0.57 per 1,000 gallons ($0.15 per 1,000 liters) for conventional treatment. Of course, the cost of ocean water treatment must be compared to the total cost of providing water from another source, which includes building dams and pipelines, installing pumps and so forth.

136. I've heard about towing icebergs to areas that are short of water as a source for drinking water. Would that really work?

Whenever we are faced with water shortages, lots of obscure ideas are floated for consideration. This is one such idea that has been considered in various forms since the 1970s, but it's never been done commercially because it would be so expensive. But it could work. Even though icebergs are floating in salt water, the ice has very little salt in it—it's basically compressed snow. A melted iceberg could provide potable water, but it would still require treatment.

137. Why are rivers dammed to create reservoirs? Can't the utility just draw what it needs from the flowing river or underground aquifers? What about the environmental impacts?

Water systems need to have enough water stored to meet the demands of their users, including residential, industrial, commercial and public use, such as fire control. If a river is running low, say in the fall or during a drought, the water in a reservoir can be drawn on to ensure adequate supplies.

Environmental impact assessments must be done for any new dams or reservoirs. This process includes evaluating the effects on wildlife and their habitats in accordance with the Endangered Species Act and other regulations and assessing instream flows of affected rivers to ensure adequate water for existing aquatic life.

138. Can wastewater be treated to make it into drinking water?

Yes. This practice, called *potable reuse*, is uncommon, but is gaining more and more acceptance around the world because of water supply limitations. The water industry commonly uses the terms *reclaimed* wastewater and *reuse* for nonpotable purposes such as park irrigation, which saves fresh water for drinking and bathing. Wastewater is already thoroughly cleaned before it is discharged to the environment, and when it is reclaimed for reuse, it is cleaned again. When it is treated for drinking water, it is cleaned again and again.

Think about it; most communities that draw their drinking water from rivers or streams already are receiving reused water from an upstream community that discharges its treated wastewater into the same body of water. And, of course, nature reuses water through the water cycle.

139. Is climate change affecting our water supply?

Water industry researchers tell us that climate change has significant implications for water supply management because increasing temperatures will increase evaporation rates, change precipitation patterns, affect runoff and water quality and change water-use demands.

Climate change has already had some effect on our water supply by increasing salinity in coastal aquifers and causing earlier snowmelt in the mountains, which affects supplies downstream. Other effects include changing water temperatures, which in turn affects the amount of natural organic matter in the water (such as algae) and subsequently affects water quality.

Water utilities are learning how to adapt to the challenges of climate change with processes such as water banking and reclamation, but we can all help preserve our precious freshwater by conserving when we can and protecting the sources from pollution.

QUALITY

140. Which is more polluted, groundwater or surface water?

It depends on what you call pollution. Because surface water can be contaminated by municipal sewage, industrial discharges, transportation accidents and rainfall runoff, it contains many pollutants but not much of any one chemical. Groundwater, on the other hand, may contain pollutants such as arsenic, nitrates, radioactive materials and comparatively high amounts of a few organic chemicals such as cleaning fluid or gasoline. Therefore, both may be polluted but in different ways.

Another difference is that the degree of pollution may change rapidly in surface water, while pollution levels change very slowly in groundwater. Your water supplier can tell you what contaminants are found in your source water, but the quality of treated water is tested more frequently than the quality of source water.

141. In towns and cities, what is the major cause of drinking water source pollution?

The major source of pollution is rainwater that flows into street catch basins (called urban runoff or stormwater runoff). While this rainwater alone is not necessarily harmful, it can pick up untreated waste products from our streets and yards, then dump it directly into rivers, streams and lakes (drinking water sources).

142. Is urban runoff treated before being discharged into drinking water sources?

Because of concerns about pollution, as mentioned in the previous question, more and more communities are capturing and treating their stormwater before it reaches drinking water sources. One way is to hold the water in detention ponds so some of the sediment and associated pollutants settle out before it is discharged. A number of manufactured devices are also available to treat runoff, including catch-basin inserts that capture debris and filter out some pollutants, such as oil; hydrodynamic separators that remove pollutants from the water through filtration or other means; and underground storage units. Such practices are expensive and can treat only a portion of the runoff.

Some communities, especially older cities, have combined sewer systems (the sanitary sewer and the storm sewer are the same), so all water is routed to a wastewater treatment plant. As cities grow, however, the combined sewer systems become too small for all the water and can overflow during a heavy storm, releasing untreated sewage to the environment. As a result, many cities are working on creating a separate storm sewer system that will capture and treat the water that flows into the storm sewer before it's discharged.

143. We used to hear a lot about acid rain. What is acid rain? Is it still a problem, and does it affect water supplies?

Acid rain forms when particles from natural sources, such as volcanoes and decaying vegetation, and man-made sources, primarily sulfur dioxide and nitrogen oxides, are released into the atmosphere. Most emissions come from burning fossil fuels such as coal to generate electrical power and gas-burning vehicles. The rain falls on surface waters and land that runs into surface waters, increasing acidity in the water. Natural aquatic and plant life have a hard time surviving in these conditions.

Because of the Clean Air Act and other environmental regulations, these types of emissions have declined in recent years and acid rain is not the threat it used to be, but it does remain in some areas.

Scientists have predicted that further emission reductions are necessary to prevent increasing damage to North American ecosystems, particularly in New England, some parts of the Rocky Mountains and much of Eastern Canada.

Water treatment plants can adjust the pH levels and balance the water between acid and alkaline, so it isn't a problem with the water you receive at the tap.

144. Do oil spills pollute drinking water sources?

Oil spilled in the oceans is harmful to the environment, but it is not a danger to drinking water sources, unless you live on the coast and your water utility is one of the few that has begun desalinating seawater. Oil spills from ship and barge accidents on lakes and rivers do contaminate surface water sources, however. Many highways and railroad tracks pass over or near drinking water sources, creating the potential for major contamination if a tanker truck or freight train accident occurs. A motor vehicle accident or improper disposal of oil from your car also can cause oil pollution. Drinking water contaminated with even a little bit of oil has such a bad taste that most people regard it as undrinkable.

Groundwater sources usually are not directly affected by oil spills, unless they occur in the recharge area, where water seeps downward to add water to the underground aquifer. Major oil spills in these areas can cause nearly irreversible damage to groundwater.

145. I live downstream from a nuclear power plant. Should I worry about radioactivity in my drinking water?

It depends. Nuclear power plants operate under strict guidelines from the Nuclear Regulatory Commission (NRC), and most operate safely to prevent dangerous radioactivity from getting into the water. However, by NRC's own admission, "a nuclear power plant may deviate from normal operation with a spill or leak of liquid material."[1] In particular, "several instances of abnormal releases of liquid tritium from several nuclear power plants ... resulted in groundwater contamination" in recent years, according to NRC. However, tritium occurs naturally in the environment and the foods we eat and disperses quickly once it enters the body, and the health risk of drinking it in levels found in some of the contaminated groundwater is considered much less than that of the radiation exposure from many medical procedures.

Also, water systems monitor the radioactive content of source water to ensure that excessive amounts are eliminated before it becomes tap water.

---

1   Fact Sheet on Tritium, Radiation Protection Limits, and Drinking Water Standards, U.S. Nuclear Regulatory Commission, http://www.nrc.gov/reading-rm/doc-collections/fact-sheets/tritium-radiation-fs.html

If you have a private well, however, you should have it tested for possible contamination.

146. How can I help prevent pollution of drinking water sources?

A lot of water contamination occurs when people or companies improperly dump solvents, cleaners, oils, pharmaceuticals and other chemicals in the ground, down storm sewers or in septic and wastewater systems. Much of this pollution cannot be traced to a single source, so USEPA calls this nonpoint source pollution.

USEPA recommends

✓ Keeping litter, pet wastes, leaves and debris out of street gutters and storm drains.

✓ Applying lawn and garden chemicals sparingly and according to directions.

✓ Disposing of used oil, antifreeze, paints and other household chemicals properly, not in storm sewers or drains. If your community does not already have a program for collecting household hazardous wastes, ask your local government to establish one.

✓ Cleaning up spilled brake fluid, oil, grease and antifreeze. Do not hose them into the street where they can eventually reach local streams and lakes.

✓ Controlling soil erosion on your property by planting ground cover and stabilizing erosion-prone areas.

✓ Encouraging local government officials to develop erosion and sediment control ordinances.

✓ Having your septic system inspected and pumped, at a minimum, every 3 to 5 years so that it operates properly.

✓ Purchasing household detergents and cleaners that are low in phosphorus to reduce the amount of nutrients discharged into our lakes, streams and coastal waters.

Remember, your drain is an entrance to your wastewater disposal system and eventually to a drinking water source. Discharges from septic tank drain fields may pollute groundwater. Treat your wastewater system with respect.

More information can be found at EPA's nonpoint source pollution Web site at www.epa.gov/nps; The Groundwater Foundation, www.groundwater.org; and Local Government Environmental Assistance Network, www.lgean.org.

147. I have a private well. Where can I get my water tested, and what should it be tested for?

Well owners should have their water tested annually for nitrates, total coliform bacteria, pH and total dissolved solids. In addition, if the well or the home is new to the owner, the water should be tested for radon and arsenic. Well owners should also ask a local agency such as a health or agricultural department about any specific contaminants of concern in their local area (e.g., pesticides or fertilizers) that should be included in the testing. These agencies will also help arrange to have your water tested, or they will explain how to take a sample for microbes and where you can take the sample for testing.

Larger cities have commercial laboratories that will test drinking water for chemicals, but the tests are more expensive and often difficult to understand. The cost will depend on the type of tests you have done. A test for microbes can be done for less than $10, but testing for chemicals can run into hundreds of dollars.

The EPA has established a Web site for private well owners, www.epa.gov/safewater/privatewells, that has a lot of good information about how to protect your well, links to public agencies by region and other materials. The Water Systems Council also has a free information service for well owners at www.watersystems council.org/network.php.

148. How can I protect my private water supply?

You can protect a private water supply by carefully managing activities near the water source, including following these recommendations from USEPA.

✓ Have your septic tank pumped and system inspected regularly.

✓ Use fertilizers sparingly and sweep up driveways, sidewalks and roads after use.

✓ Never dump anything down storm drains or onto the ground around the well, house or underground piping.

✓ Revegetate or mulch disturbed soil as soon as possible.

✓ Clean up spills of vehicle fluids or household chemicals and properly dispose of cleanup materials.

✓ Minimize pesticide use and learn about integrated pest management.

✓ Direct roof drains away from paved surfaces and bare soil.

✓ Take your car to a car wash instead of washing it in the driveway.

✓ Check your car for leaks and recycle motor oil.

✓ Pick up after your pet.

Distribution

When you put your hand in a flowing stream, you touch the last that has gone before and the first of what is still to come.
— Leonardo da Vinci

149. Are the pipes that carry drinking water from the treatment plant to my home clean?

A well-run water utility will have a program of flushing and cleaning the distribution pipes to ensure that the water delivered by the distribution system is safe. Water distribution pipes made of metal or concrete can corrode internally because of aggressive water quality characteristics that react with the inner surface of the pipe. Generally, corrosion by-products are not harmful to human health, but they can cause nuisance issues, such as spotting on laundry or a metallic taste in the water.

If chemical reactions in the water cause sulfates and carbonates to form, the sulfates and carbonates can stick to the walls of the pipe and form scale. Scale can clog fixtures and reduce the flow of water in pipes. Periodic water pipe flushing keeps the buildup of corrosion by-products and scale to a minimum. How often a water system needs flushing depends on water chemistry and pipe material.

To flush pipes, water department employees close valves and open fire hydrants in a designated sequence, forcing water pressure to concentrate in certain pipes before the water rushes out of a hydrant, along with the buildup from the pipes. Alternatively, pipes may be cleaned by using water pressure to force a tight-fitting plastic swab through the pipe. The swab, sometimes called a *pig*, scrapes the pipe walls clean. Corrosion by-products and scale are then flushed out of the system through an open fire hydrant. Similar pipe-cleaning devices with other designs are also available.

Keeping water pipes clean is a big job, as there are about 1 million miles (1.7 million kilometers) of pipes in the United States and Canada.

150. I have seen work crews cleaning water mains, and the water they flush out looks terrible. How can the water be safe if the pipes are so dirty?

As discussed in the previous answer, water suppliers' regular flushing and cleaning programs remove corrosion and scale from the walls of several miles of pipe. When this material flows out of a fire hydrant all at once, it looks worse than it really is. If you watch the workers do this, you will notice that the water clears up rather quickly. Then the flushed system is disinfected, and samples are taken to test for bacteria; the flushed pipe is not put back into service until the water tests are clean.

151. What are distribution pipes made from?

Pipe may be made of ductile iron, steel, concrete, poly-vinyl chloride (PVC) or high-density polyethylene (HDPE). Ductile iron, steel and concrete pipes are strong and durable and can be used in systems with very high pressures. PVC pipe is light, inert to most chemicals and noncorrosive. It is easy to install and can be cut with a hand saw. HDPE pipe is softer and more bendable than PVC and is suitable for lower pressures and tighter turning radii applications.

152. Why are fire hydrants sometimes called fire plugs?

Long ago, tap water was distributed in hollowed-out log pipes. When water was needed to fight a fire, the pipe was dug up and a hole was drilled in the wooden pipe. When the fire was over, a wooden plug was used to close the hole. These fire plugs were then marked for possible future use.

153. We don't use much water for drinking. Why does all the rest need to be treated so extensively? This seems unnecessarily expensive.

Actually, the cost of water treatment itself is a small part of your water bill—only about $1.50 to $3 each month for each person living in your house—but that is not the real answer. An additional distribution system could be developed in which two pipes come from the treatment plant, through the streets and into your home. This dual distribution system would feature a small pipe for drinking water and a larger pipe for all of the other water uses (toilet flushing, lawn watering and so forth). To take out all of the pipes and install a dual system in existing cities would cost a lot and would disrupt services during the conversion. However, many communities are now recognizing the value of reclaimed or reuse water as a way to satisfy water demands, particularly for uses that do not require potable-quality water, and are modifying their systems in this manner. In any case, the water provided for purposes other than drinking still needs to be treated to ensure minimum health risk.

154. How does a water utility detect a major leak in the distribution piping system?

A major leak can be detected by

✓ Visual observation (water on the ground or spraying water)
✓ A loss in water pressure
✓ A depression in the ground or road over a water main
✓ Reports by public-minded citizens
✓ Sensitive listening devices that detect the sound of the leaking water underground

Also, in metered systems, water utilities regularly compare the amount of water produced to the amount that passes through customer meters. This method provides an excellent accountability of overall water loss.

Stopping leaks is important to water suppliers because leaks waste water, adding costs to both the water supplier and you. Water utilities don't get paid for the water that is lost to leaks, but the cost for treatment and distribution must still be passed along to customers. The national average for unaccounted water from all sources is 15 percent of treated water, although many suppliers keep such losses lower. Gas companies only lose about 5 percent of their product.

The water utility is responsible for the underground pipes from the street up to the water meter, but if your pipes leak within the boundary of your property after the meter, it is your responsibility to repair them. Prompt repair is to your benefit: a leaking pipe means a higher water bill.

155. If leaks are such a waste of water, why can't the water utility just fix all of them?

Most pipes in a municipal water distribution system were laid before many of us were born and are failing at different rates, depending on the type of pipe, ground conditions, usage and other factors. Current generations have not had to pay huge amounts to repair and replace these pipes as they age and fail. To repair the miles of aging pipes is a massive job, not to mention the cost. In fact, much of the drinking water infrastructure in the United States does need to be replaced in the next three decades. The real issue is how utilities are going to pay for all this repair and replacement.

The cost estimates to replace all the old pipes in the United States range from $280 billion to $400 billion. The federal and state governments offer some relief to utilities to offset the cost of repairs, generally in the form of low-interest loans, but in the end, water rates will have to rise to pay for repairing and replacing current water infrastructure.

156. Fixing a broken water pipe looks like a dirty job. How is the inside of the pipe cleaned afterward?

After work is done, the pipe is flushed to clear it of loose debris that may have collected during the repair. Then the pipe is filled with water containing a disinfectant— usually chlorine or a derivative. Holding this water in the pipe for a time kills bacteria and other organic contaminants in the system. This is followed by testing to be sure the water is safe.

This is not the end of the story, however. The utility must take care in disposing of all this water that contains so much chlorine. State, provincial and federal regulations control its disposal. A chemical must be added to inactivate the chlorine before the water can be flushed out of the pipe and discharged, unless the highly chlorinated water is discharged to a wastewater treatment plant or another location where it will not have an adverse impact on the environment.

157. What causes low water pressure?

Temporary low pressure may be caused by heavy water use in your area—lawn watering, a water main break, fighting a nearby fire and so on.

Permanent low pressure could be caused by the location of your home (on a hill or far from the pumping plant) or pipes that are too small. If you have an older home, the pipes in your home could have a lot of scale in them, leaving little room for the water to flow. Sometimes water pressure problems are as simple as a plugged faucet aerator, and sometimes the design of the interior plumbing can cause low pressure.

Low pressure is more than just a nuisance. The water system depends on pressure to keep out contamination and to ensure enough flow to fight fires. If the pressure drops, the possibility of contaminants entering the drinking water increases. One of the causes of poor quality water in some developing countries is low pressure in the distribution system that allows contamination to enter the pipes. You should report any permanent drop in water pressure to your water company.

Many areas have minimum standards for pressure. For example, 20 pounds per square inch (psi)—140 kilopascals (kPa)—when water use is at a maximum is a common standard (car tires often use 30 to 32 psi of air [about 210 kPa]). Most systems have pressures three to four times the minimum.

You can tell you may have low pressure if flows from your faucets at home are much lower than elsewhere in your area: at work, in a restaurant washroom or in a friend's home elsewhere in the city, for example.

The only way to cure permanent low pressure is to have the supplier change the system, adding more pumps or bigger lines. If the problem is in your home (more of a nuisance than a potential health hazard), check the faucet aerators for any accumulated sediment and discuss your options with a reputable plumber. For example, there may be a bottleneck in your piping, valves or water meter that could be easily corrected. Or, if you have newer fixtures, they may be designed to restrict flow as a conservation measure.

You may be surprised to learn that you can also have too much pressure. Some homes need pressure regulators to avoid damaging household plumbing from very high water pressures.

158. What are cross-connections?

A cross-connection is a connection between a drinking water pipe and a contamination source. Here's a common example: You're planning to spray weed killer on your lawn. You hook up your hose to the faucet on your house and to the sprayer containing the weed killer. If the water pressure drops at the same time you turn on the hose, the chemical in the sprayer may be sucked back into your home's plumbing system through the hose. This is called *backsiphonage* and would seriously contaminate the water system in your home. If your hose was connected to a fire hydrant or a public access faucet (e.g., at a campground), then the weed killer would be sucked into the public water supply.

Backsiphonage can be prevented by using an attachment on your hose called a backflow-prevention device. The simplest backflow-prevention system is an air gap, which is a physical separation of the supply pipe by at least two pipe diameters vertically above the overflow rim of the receiving vessel (the sprayer containing the weed killer in the example). A hose bibb vacuum breaker installed on the outdoor spigot will also work.

More sophisticated backflow-prevention devices are mandatory for certain industrial and commercial operations, such as dry cleaners and restaurants. Most water suppliers have cross-connection control programs, particularly in major cities.

159. Why is some drinking water stored in large tanks high above the ground?

Aboveground storage ensures that water pressure is fairly constant and water volumes are sufficient to fight fires, even if the electricity that runs the water pumps is off. The storage tanks also provide a steady supply of drinking water when water use is high. The tanks are usually filled late at night when drinking water use is low. Water suppliers must pay attention to the operation and maintenance of storage tanks so the water circulates frequently and stays fresh.

Conservation

When the well's dry, we know the worth of water.
— Benjamin Franklin, *Poor Richard's Almanac,* 1746

160. What indoor home activity uses the most water?

Inside the home, water use is evenly distributed among appliances, but nearly 30 percent is flushed down the toilet. According to the U.S. Environmental Protection Agency (USEPA), a typical household of four uses 400 gallons of water per day. Clothes washing accounts for 26 percent of that use, followed by showers at 20 percent and faucets (washing dishes, brushing teeth, etc.) at 19 percent.

To put this in perspective, a person needs only five gallons of clean water a day to meet basic needs, according to the World Health Organization. Many people in undeveloped countries do not have access to even that much clean water.

161. How much water can we save by installing new fixtures?

Water-efficient toilets and bathroom faucets or faucet accessories can save the average home more than 11,000 gallons (41,600 liters) per year, according to USEPA. Low-water-use toilets require less than 1.3 gallons (4.9 liters) per flush, compared to 3.5 gallons (13.2 liters) per flush for older, less efficient models. If you replace your old clothes washer with a high-efficiency washing machine, you can reduce water consumption from 40 gallons per load to less than 28 gallons (106 liters) per load. And if one out of every 10 U.S. homes upgraded to water-efficient fixtures, more than 120 billion gallons (454 billion liters) of water and more than $800 million would be saved annually. You can find out about water-saving appliances by consulting WaterSense, an USEPA program that provides information on products and programs that save water without sacrificing performance, online at www.epa.gov/watersense.

Using less water conserves in other ways as well. It takes a considerable amount of energy to treat and deliver the water you use every day, and USEPA has determined that if just one out of every 100 American homes were retrofitted with water-efficient fixtures, about 100 million kilowatts per hour of electricity could be saved per year—avoiding 80,000 tons (72,600 metric tons) of greenhouse gas emissions. That is equivalent to removing nearly 15,000 automobiles from the road for one year! And it will help your electric bill, too. Letting your faucet run for five minutes uses about as much energy as letting a 60-watt light bulb run for 14 hours.[1]

162. Why can't I just put a brick in my toilet instead of replacing it?

Putting something in the toilet tank that takes up space means that less water is needed to refill the tank after a flush—and that is a good idea—but putting a brick in your toilet tank is not smart. A brick tends to crumble

---

1   U.S. Environmental Protection Agency. 2009. Benefits of Water Efficiency. WaterSense Web site: www.epa.gov/watersense/water/benefits.htm

and might damage the toilet's flushing mechanism or clog the water inlets. Instead, use a plastic jug filled with water. Because some toilets require a certain volume of water to work properly, be sure to test the toilet to make sure it's still flushing well after any changes.

163. Before I replace my showerhead, how do I measure how fast my shower is using water?

First, know that the water coming from a low-flow showerhead should not exceed 2.5 gallons (9.5 liters) per minute. Next, get a bucket and a stopwatch or a watch with a second hand. Make sure the bucket has a 1-gallon (3.8-liter) mark on it. If it doesn't, add a gallon of water and mark the level. Set the shower flow just as you would when showering. Catch all the water in the bucket for 24 seconds. If after that time, the water level is near the 1-gallon mark, your showerhead is flowing at the recommended amount. If the level is over the 1-gallon mark, replace your showerhead with a new low-flow model.

164. What about outside use— how can I conserve water there?

At least 30 percent of water used by a single-family suburban household is for outdoor irrigation, and much of that goes to waste through evaporation or runoff caused by overwatering. To conserve water, consider these landscaping tips

✓ Maintain a lawn height of 2 ½ to 3 inches to help protect the roots from heat stress and reduce the loss of moisture to evaporation.
✓ Promote deep root growth through proper watering, aerating, fertilization, grass-clipping control and attention to lawn height. A lawn with deep roots

requires less water and is more resistant to drought and disease.

✓ Avoid planting turf in areas that are difficult to irrigate properly, such as steep inclines and isolated strips along sidewalks and driveways.

✓ Aerate clay soils at least once a year to help the soil retain moisture.

✓ Mulch around plants, bushes and trees to help the soil retain moisture, discourage the growth of weeds and provide essential nutrients.

✓ Plant in the spring or fall, when watering requirements are lower.

✓ When choosing plants, keep in mind that smaller ones require less water to become established.

For other outdoor uses, consider these tips

✓ Use porous materials for walkways and patios to keep water in your yard and prevent wasteful runoff.

✓ Use an adjustable hose nozzle for plant irrigation, and be sure to shut water off at the house connection each time the exterior faucet is used.

✓ Use a broom or rake to remove debris from driveways and walkways, not water.

✓ If you have a pool, keep the water level low to minimize splashing, and use a cover to slow evaporation. An average-sized pool can lose about 1,000 gallons (3,785 liters) of water per month if left uncovered.

✓ Use a bucket of soapy water and a hose with a shut-off nozzle to wash your car.

165. How should I irrigate my lawn and garden to avoid wasting water?

Water early in the morning to avoid excessive evaporation from sun and wind; water pressure is typically higher then, too. Water your lawn for extended periods a couple of times each week, rather than every day, to allow deep moisture penetration, unless you have dense, clay-like soil that does not absorb water readily. If this is the case, split your watering time into shorter periods to allow for better absorption. An inch of water a week is a good rule of thumb, but varies for different grasses and different parts of the country.

Before watering, use a trowel to check the root zone of your lawn or garden for moisture; if the soil is still moist two inches below the surface, you don't need to water. Or install an automatic irrigation sensor that measures the moisture content of the soil and controls when and where an area is watered. This is particularly helpful to prevent watering when it has just rained.

For a real difference, replace your sprinkler system with a drip irrigation system. According to EPA, a drip system uses between 20 to 50 percent less water than conventional in-ground sprinkler systems, and no water is lost to wind, runoff and evaporation.

166. Can I save water by changing my landscaping to Xeriscape?

The majority of landscape problems are directly related to overwatering. Xeriscape is increasing as a trend in landscape design that combines water conservation practices with creative landscape design that emphasizes the use of plants that are appropriate for your climate and soil conditions. A properly designed and operated irrigation system can reduce water use by 20 percent or more each year and ensure a better-looking yard and garden.

Xeriscape calls for grouping plants according to their watering needs, along with proper soil preparation, using shade, rethinking traditional grass lawns, taking advantage of natural runoff, planting in low irrigation areas and using mulch.

167. I heard that some water conservation measures are actually required by law. Is this true?

In the United States, low-volume toilets and shower fixtures have been required since 1994 in new homes and businesses, and since 1997 when fixtures are replaced, according to the Energy Policy Act of 1992. This law has helped fund the toilet rebate programs that many big-city water utilities offer. Some states, led by California, have adopted water-efficient landscaping regulations and other restrictions on water use.

Canada has made significant efforts to introduce requirements and programs encouraging water conservation through water-efficient appliances and fixtures. Ontario, in particular, has initiated low-flow toilet installation programs. Limitations on water supply, as well as limitations in the capacity of utility infrastructures, will eventually require utilities everywhere to introduce water saving measures, and, if the public doesn't willingly participate in conserving water, more regulations may be required to protect this vital resource.

168. Which uses more water, a bath or a shower?

A full bathtub requires about 70 gallons (265 liters) of water, while taking a five-minute shower under a low-flow showerhead uses 10 to 25 gallons (38 to 95 liters). If you do take a bath, don't run water down the drain while it heats up; adjust the temperature as you fill the tub.

169. I leave the water running while I brush my teeth. Does this waste much water?

You bet! Leaving the water running is a bad habit; about 4 to 6 gallons (20 to 25 liters) of water go down the drain needlessly every time you brush because the average bathroom faucet runs at a rate of about 2 gallons (7.5 liters) a minute. Turning off the water when you are not using it will save water and save you money.

170. I use a lot of water in the kitchen. How can I conserve there?

✓ Remove residue from each cooking utensil and dish without using water and don't rinse them before putting in the dishwasher.
✓ Clean vegetables in a pan of water rather than under running tap water, then use that water to give your plants a drink.
✓ Use the garbage disposal sparingly.
✓ Run the dishwasher only when it is full.
✓ Keep a container of water in the refrigerator instead of running the tap to cool your drinking water.
✓ Use a pail or basin instead of running water for household cleaning. A sponge mop will use less water than a string mop.
✓ Presoak grills or oven parts overnight when they need cleaning. Wash with an abrasive scrub brush or pad and use plenty of "elbow grease" to minimize water use.

171. Many water quality problems in the home—lead, red water, sand in the system and so forth—are cured by flushing the system. Isn't that a waste of water?

Yes, but you can avoid losing this water by catching it in a container and using it for plant and garden watering. Flushing water pipes keeps lead or rust out of the water, and that benefit will outweigh the loss of a few gallons. Try to use your flush water, but if you can't, don't feel too bad. This water has served a useful purpose.

172. My water faucet drips. Should I bother to fix it?

Yes. Drips waste precious water, and you will pay for it eventually. To find out how much water is wasted, place an 8-ounce (23.6-centiliter) measuring cup (or anything that will let you measure 8 ounces) under the drip and watch how many minutes it takes to fill it up. Divide the filling time into 90 to get the gallons of water wasted each day. For example, if your faucet fills the measuring cup in less than 30 minutes, it could be dripping 60 times a minute (once each second), adding up to more than 3 gallons (12 liters) each day or 1,225 gallons (4,630 liters) each year, just from one dripping faucet.

173. How concerned should I be about a leaky toilet?

According to the EPA, 11,000 gallons of water every year is lost from the average American household because of running toilets, dripping faucets, and other household leaks. A leaky toilet can waste as much as 200 gallons (757 liters) of water a day. A common reason toilets leak is that the toilet flapper has become worn and no longer seals closed once the toilet has filled. Flappers are inexpensive rubber parts that can build up minerals or decay over time, particularly if you use a chlorine tablet in the tank. If you are handy, it's easy to replace the flapper, and advice is available on the Internet. For more information, check out "Fixing Leaks Around the Home," EPA's consumer Web site, at www.epa.gov/watersense/fixaleak/howto.htm.

174. I have a private well. Why should I conserve water?

Many aquifers in this country are seriously depleted, and groundwater sources are in peril. By conserving water, you can help maintain the source and avoid having to have your well drilled deeper to reach the aquifer.

175. Why do we still have a water shortage when it's been raining at my house?

It takes weeks to years for the water to recycle from rain to home use, depending on your water source. If your drinking water comes from a well, the rainwater needs to recharge into the ground before it affects the supply. If it comes from a reservoir or surface source, the supply may be far upstream from the rain at your house, so the water source isn't being replenished. Although the rain in your area may not help the water supply, it does lower the demand for water for watering lawns and gardens and washing cars.

176. During water shortages, shouldn't decorative fountains be turned off?

In most cases, fountain water is recirculated (used over and over) and is not wasted. If water losses from evaporation are high, however, fountains should be turned off.

177. Do restaurants actually conserve water by not serving it to customers?

Skipping water in restaurants serves as a good reminder to everyone about the importance of saving water, but the actual volume of water saved is small. The water that would be used to wash the water glasses is also saved, and this is usually more than the glass of drinking water served to customers (two glasses of water for each glass washed).

However, health experts emphasize the importance of drinking at least six to eight glasses of water each day, so if you want water to drink, ask for it. Skip it if you are just going to let the glass sit on the table while you drink something else.

178. Since the amount of water on the globe isn't changing, and the water in my area is plentiful, why should I conserve?

You're right about the amount of water being constant, but conservation is still important. For example, suppose you live in a growing community. As the population increases, so does the demand for water. This means that every so often, the water supplier must find another source of water, and in some areas additional sources are hard to find. If everyone conserved, the water demand would not grow as fast, and the need to look for more water would be delayed. This permits the municipality to defer expenditures and to use the money for something else in the meantime. In addition, not all of the water taken from the tap returns immediately to the source. Aquifers, for example, take a long time to recharge and may never come back to their full capacity. You can also look at it this way: when we use water, we generally add contaminants to it. Even though the amount of water on the planet is constant, if we use the water for drinking, cleaning and other domestic, commercial and industrial purposes, these contaminants still must be removed.

179. What is gray water? Can we use it?

Gray water is used water from dishwashing, showers, sink use or laundry water in the home—essentially any residential wastewater other than water from toilets. Gray water can be reused for other purposes, especially landscape irrigation because gray water can contain small bits of soil-enriching residue.

According to Oasis Design's Gray Water Policy Center, the benefits of using gray water include

✓ Less use of fresh water
✓ Less strain on septic tanks
✓ Less energy and chemical use
✓ Groundwater recharge
✓ Plant growth
✓ Reclamation of otherwise wasted nutrients[2]

Note that in some communities, particularly in Western United States, capturing gray water is not allowed.

180. Our water utility has opened a reclaimed water plant in our town. What is reclaimed water? Can we drink it?

Reclaimed water, also referred to as *reuse water* or *recycled water*, is wastewater treated to reusable standards, but it is not intended as a drinking water source. Water

---

2   Oasis Design. 2009. Gray Water Policy Center. www.oasisdesign.net

users, whether industrial, commercial or municipal, are increasingly being supplied with reclaimed wastewater in place of potable water because it costs substantially less than developing new high-quality water sources for potable purposes. Reclaimed water is made suitable and safe for reuse through extensive wastewater treatment and by limiting public or worker exposure to the water through design and operational controls.

This water is distributed through a separate piping system—you may have seen the purple pipes being installed—and is primarily used for irrigation of public parks, cemeteries, plant nurseries, medians and, if the system is reconfigured for hydrants, firefighting. Other uses may include industrial cooling and heating, toilet and urinal flushing, car washes and other services.

181. Is reclaimed water regulated?

USEPA published guidelines for water reuse in 2004, but there are no federal regulations governing water reclamation and reuse in the United States. Several states, however, have established regulations or criteria regarding this water. Do be assured that any water that is being reused for human use does have to meet minimum safety standards and is cleaner than the water that is discharged directly from a wastewater plant into the environment.

If wastewater can be used productively, the savings in wastewater treatment can be passed on to users, making water reclamation and dual distribution systems more economically attractive.

Regulations, Reporting and Water Security

Providing safe drinking water and sanitation for all cannot be merely a slogan but rather a call to action so that future generations will have a drop to drink.
— Jack Hoffbuhr, former executive director of American Water Works Association

182. Is the quality of drinking water regulated by legislation?

In the United States, the Safe Drinking Water Act (SDWA), administered by the U.S. Environmental Protection Agency (USEPA), was first passed in 1974 and expanded and strengthened in 1986 and 1996. The SDWA protects the quality of drinking water by establishing health-based limits on foreign matter in water, limiting everything from man-made chemicals and disease-causing microbial contaminants to water additives, such as fluoride and chlorine.

In Canada, drinking water is also a shared responsibility of provincial, territorial, federal and municipal governments. The federal government sets water quality guidelines and regulates federal facilities. The provinces and territories generally are responsible for regulating the provision of safe drinking water to the public, while municipalities usually deliver water to the public. Health Canada uses its scientific and technical expertise to develop *Guidelines for Canadian Drinking Water Quality* in partnership with the provinces and territories. These guidelines are used by every jurisdiction in Canada and are the basis for establishing drinking water quality requirements for all Canadians.

183. What about state or provincial regulations?

In most U.S. states, the health department or department of environmental quality has the primary responsibility for enforcing federal standards. Each state must adopt drinking water quality standards that are at least as strict as those of the federal government. The responsible agency evaluates its own public water systems and rates them as in compliance or out of compliance, and takes action to ensure those out of compliance improve their operations to ensure safe drinking water for their customers. In Canada, a provincial agency, typically the department of health, environment or municipal affairs, establishes regulations and guidelines pertinent to that province. The provinces have more control than the federal agency in regulating the water in Canada.

184. Is water that meets government drinking water standards absolutely safe?

Safety is relative, not absolute. For example, an aspirin or two may help a headache, but if you took a whole bottle at once, you'd probably die. So, is aspirin safe? When setting drinking water standards, regulatory agencies use the concept of reasonable risk, not risk-free. Risk-free water is not technologically feasible and would cost too much. So the answer to the question is no, drinking water isn't absolutely safe. But the likelihood of getting sick from drinking water that meets the federal standards is very small, typically one chance in a million.

One difficulty federal agencies have when trying to determine reasonable risk relates to people who are more likely to get sick than others, called the *susceptible population*. For example, only babies three months old or younger are seriously affected by nitrates in drinking water, so for that contaminant they are the susceptible population. The standard for nitrate in drinking water, therefore, was chosen to protect infants. With other contaminants, identifying the susceptible population is not as easy. Are they babies, elderly, undergoing cancer treatment, in nursing homes, HIV positive, or others? For each standard, the federal regulatory agencies must balance the risk to all these groups against the cost of treatment and arrive at a standard that will protect as many people as possible at an affordable cost. This is called "the greatest good for the greatest number."

185. How do regulatory agencies choose the standard for a chemical in drinking water?

In the United States, the National Institute of Environmental Health Sciences through the National Toxicology Program and USEPA laboratories conduct extensive tests on rodents with the chemical in question to determine its effects. Because rats and mice digest their food in the same way humans do, this information is extrapolated to determine a *reasonable risk* of exposure to the chemical over time. For most potentially cancer-causing chemicals, reasonable risk is defined as follows: If 1 million people drank water for a period of 70 years with the amount of chemical in it equal to the standard, no more than one additional person would be likely to get cancer from the drinking water—a very small risk. If possible, a safety factor is added that lowers the allowable exposure even further to determine the drinking water standard.

The federal agencies also conduct studies on appropriate treatment and monitoring technologies and work with water systems to collect contaminant occurrence data, which are used to conduct a cost–benefit analysis that both determines if the contaminant must be managed and what management actions are cost-effective. This provides the appropriate level of public health protection given the nature and distribution of the contaminant.

In Canada, the Federal–Provincial–Territorial Committee on Drinking Water (CDW) has been charged with of identifying, assessing and evaluating potential contaminants. CDW uses risk assessments from both USEPA and the World Health Organization for establishing standards for national guidelines, along with regional research information on the occurrence of specific parameters.

186. How do we know what the allowable level of a contaminant is?

Once a drinking water standard has been chosen for any contaminant, it is called the *maximum contaminant level*, abbreviated MCL, for that contaminant. In addition to the MCL, by law, USEPA must choose a maximum contaminant level goal (MCLG) for any regulated contaminant. This is the level below which there is no known or expected risk to health. Contrast this with the reasonable risk concept. The MCLG also contains a factor of safety. USEPA does not enforce MCLGs.

In Canada, the *Guidelines for Canadian Drinking Water Quality* is just that—guidelines—and unless a province adopts the guidelines as enforceable requirements, the allowable level is just a recommendation. Health Canada does set a maximum acceptable concentration (MAC) for parameters believed to have a health impact. Check with your local utility if you have concerns about the level of a given parameter in your water.

187. Most federal standards are written like this: "Selenium—0.05 mg/L." What does "mg/L" mean?

The abbreviation mg/L stands for milligrams per liter. In metric units, this is the chemical's weight (selenium in the example) dissolved in one liter of water. One liter is about equal to 1 quart, and 1 ounce is equal to about 28,500 milligrams, so 1 milligram is a very small amount. About 25 grains of sugar weigh 1 milligram.

Because 1 liter of water weighs 1 million milligrams, 1 milligram per liter (mg/L) is sometimes written 1 part (weight of chemical) per million parts (weight of water), or 1 ppm (part per million).

One part per million is hard to imagine. Here are some examples: 11.6 days contain 1 million seconds, so 1 second out of 11.6 days is 1 part per million (1 second in a million seconds), or 25 grains of sugar in a quart of water is about 1 part per million, or 1 drop in a 55-gallon drum is about 1 part per million.

188. What is the Clean Water Act?

Officially known as the Federal Water Pollution Control Act, the Clean Water Act (CWA) was established by the US government to control the treatment of wastewater and its subsequent release into the environment. Under the CWA, any facility that intends to discharge waste materials into the nation's waters must obtain a permit before initiating a discharge. The intent of the CWA is to restore and maintain the chemical, physical and biological integrity of the nation's waters by preventing pollution through the setting of standards for the treatment of discharges and providing assistance and guidance to wastewater treatment facilities and other industries that discharge to the waterways.

In Canada, the Fisheries Act protects water bodies from pollution and each province and territory develops its own legislative framework for regulating water.

189. How do regulatory agencies choose the contaminants to regulate?

USEPA has developed a Contaminant Candidate List (CCL), which identifies contaminants that are not regulated by national primary drinking water regulations under the SDWA, but are known or anticipated to occur at public water systems and may warrant regulations in the future. Certain water systems are required to monitor for certain unregulated contaminants under the Unregulated Contaminant Monitoring Rule (UCMR). USEPA then analyzes contaminant occurrence, health effects and other factors to determine if a contaminant should be regulated. Canadian standards are based on similar occurrence and health effect studies.

190. If most tap water is safe, why are engineers and scientists still doing so much research and why is the government thinking about more regulations?

Drinking water suppliers are committed to keeping water safe for consumers and do so by adhering to the tenets of the SDWA. The CCL is used to prioritize research and data collection efforts to help determine whether a specific contaminant needs to be regulated because of its health effects and/or occurrence levels. Because monitoring equipment has become so sophisticated, it can detect even the minutest portion of a foreign element in water, and medical research continues to identify causal links to various diseases. As new contaminants emerge that need to be researched, numerous organizations that focus on water research will conduct studies that will be shared with water utilities, health professionals and regulators to determine what course of action should be taken to control the contaminant.

191. Must all surface water supplies be filtered?

Since 1993, USEPA has required all U.S. water suppliers using surface water to include a filtration step during treatment. Waivers to this rule have been allowed if a water system can demonstrate exceptional water quality and show that water quality can be protected without filtering by implementing strict watershed protection measures. Since the late 1990s, however, USEPA has required some systems with filtration waivers, including those in New York, Boston and Seattle, to build filter plants. If you are interested in your local situation, talk to your water supplier.

In Canada, a similar situation exists. General policy requires filtration of surface water, but a local supplier that can demonstrate it meets stringent testing, reporting and quality criteria need not filter. As in the United States, many localities are assessing the need for filtration.

192. What is the Ground Water Rule?

The U.S. Ground Water Rule (GWR) was enacted in 2006 to protect the public from pathogen contamination in public drinking water systems that use groundwater. Historically, groundwater has been considered to be pathogen free, but some groundwater sources are contaminated and treatment is necessary. The GWR applies to all systems using groundwater, but it does not require all groundwater systems to treat their water. Under the rule, treatment plants must identify any problems that could cause contamination, which may then lead to a requirement for disinfection.

193. How is my water tested and who tests it?

All public water systems in the United States and Canada must test the treated water for nearly 100 parameters a specified number of times each year. The tests for microbes are conducted most often; the frequency varies depending on the population served by a water supplier. Rather than test for the actual pathogens, water systems test regularly for *indicator* microbes that are more readily collected and analyzed, and which provide a good measure of whether the system is adequately protecting public health.

In the United States, these tests are conducted in state certified laboratories using federally approved methods, some of which are quite complex. In Canada, the testing labs are nationally accredited and, in most cases, must be approved by the local or regional government, such as the province or First Nation. Private wells may be tested in connection with the sale of a home, but it is up to individual homeowners to test their wells on a regular basis.

194. Can water systems be excused from monitoring for some contaminants?

In some cases, yes. Over time, by reviewing test results and keeping a watchful eye on potential problems, public water systems come to understand the likely threats to their water supplies. If a water utility does not have water quality problems, it can apply to its regulator for permission to test less frequently for certain contaminants.

If regulators believe human or natural activities are unlikely to contribute that contaminant to the system's water, the request to avoid unnecessary testing may be granted. A waiver from monitoring requirements in no way reduces the supplier's responsibility to provide high-quality drinking water.

195. How do I find out if my water is safe to drink?

If your water comes from a public water system, contact your water supplier or local health department and ask if the water meets federal, state or provincial standards. If it does, the water is safe to drink. (This is a good practice when you move to a new location.)

Most water utilities in the United States are required to provide customers with an annual water quality report called a *Consumer Confidence Report*. Required by the 1996 amendments to the Safe Drinking Water Act, the report must provide customers with basic information about their drinking water, including its source; susceptibility to contamination; level or range of levels of any contaminant found in the water, the MCL for that contaminant and the likely source of that contaminant; the potential health effects of any contaminant detected that is above the MCL and the utility's actions to mitigate the contamination in tap water; and other information.

USEPA also has compiled information about local water systems on the Web at www.epa.gov/safewater/dwinfo/index.html, or you can call the Safe Drinking Water Hotline at 1-800-426-4791.

Canadian law requires water quality problems to be reported to public health agencies but does not require violations to be reported to consumers.

If you have your own private water source, such as a well or spring, you are responsible for having it tested. Once it has been tested, you can discuss the results with your local health department.

196. What information should I look for in the water quality report from my supplier?

Most reports contain a table of constituents found in the local drinking water. For each constituent, the table usually shows USEPA's maximum contaminant level (MCL) and the amount found in your drinking water. If the amount in your supply is the same as or less than the MCL, your supply is all right. This will be the case in the majority of the situations.

Note that the amounts of some constituents may be listed as "not detected" (ND) or "below detection limits" (BDL). This is not the same as zero, because a level of zero would mean that not any of that constituent was in the water—not even one molecule. Because testing instruments cannot measure that small an amount, only the smallest amount of material that will cause a reading is known. If no reading is obtained, any trace of that constituent in the sample was too small to register. The instrument operator then reports ND or BDL for that constituent.

If the MCL for some constituents is listed as "treatment technique," instead of a number, the utility is required to install and properly operate water treatment processes that reliably remove certain contaminants that are not economically or technologically feasible to measure, including *Giardia*, *Legionella* and *Cryptosporidium*.

197. I've received a notice from my water utility telling me that something is wrong. What's that all about? What is a boil-water order?

In the United States, federal law states that water suppliers must promptly inform consumers if the water has become contaminated by something that could cause immediate illness. If such a violation occurs, the supplier has 24 hours to notify the public through the media or personal contact. The notification will also include information about whether the water should be boiled (called a *boil-water order*) to kill organisms before use, whether it is safe to use for nonpotable purposes and what actions are being taken by the water supplier to solve the problem. This is also common practice in Canada.

If you or someone in your household is immunocompromised, elderly or very young, you may want to take extra precautions when there is a water alert because these people may be more vulnerable to contaminants in drinking water than the general population.

All violations are important, of course, but they are not all equally important. For instance, if the problem reported to customers is that the water supplier has not sampled the water as frequently as required by the regulations, this does not necessarily mean that the quality of the water is poor.

One violation involving insufficient sampling, improper reporting or water quality does not mean the water is unsafe to drink. It does mean, however, that the water utility should improve its operations.

Remember, if you have a boil-water order in your area, be sure to throw all of your ice cubes away. They may have been made with contaminated water.

198. I live in an apartment and don't get a water bill. How will I know if there are any problems with my tap water?

Serious problems, called *acute violations*, must be broadcast on the radio and television and published in general circulation newspapers. Less serious problems must be published in general circulation newspapers and may also be otherwise posted or be sent to everyone who pays a water bill—in your case the apartment owner or manager.

Ask your apartment manager to post anything included with the water bill. Even when no problems exist, water suppliers often include valuable information (often called *bill stuffers*) with the bill. Renters should be able to see this information.

199. What are the chances of a water system being the object of a terrorist attack?

Just like many public facilities, water systems are considered by the federal government as a possible target of terrorist activity. Because of this, the Bioterrorism Preparedness and Response Act of 2002 (Bioterrorism Act) required all community water supply systems in the United States to conduct a risk assessment to determine what facilities and operations were most vulnerable to an attack. After the assessment, many utilities took actions to protect what they identified as their most vulnerable assets, be it the treatment facility, water distribution system, computer programs or other components of the water system.

There are several federal initiatives through the Department of Homeland Security that have provided guidance to utilities in implementing security measures. Many water utilities are also continually monitoring threat activity and are part of a confidential national security alert program called the Water Information Sharing and Analysis Center (WaterISAC).

In a recent survey that resulted in a report to Congress[1], the U.S. water sector (drinking water and wastewater) appear to be making good progress in implementing security policies, procedures and training at their utilities.

---

1   Water Sector Measures Analysis, prepared for the Water Sector Coordinating Committee and WaterISAC, December 30, 2008. http://www.awwa.org/files/GovtPublicAffairs/PDF/MetricsReport.pdf

200. What would happen if somebody intentionally put a dangerous chemical in our water?

Because water utilities are aware of the potential for malevolent attacks on water supplies, most have increased their monitoring and testing of water quality as a result of possible threats. This means that if something strange was added to the water, the water quality parameters, such as pH levels, flow and turbidity (the amount of suspended particles in water), would noticeably change and water system operators would be alerted during routine testing or through automated devices that constantly measure these parameters. The operators would then conduct more testing to determine what the added substance was and take precautions to prevent the contamination from becoming a problem at the tap.

If the contamination had already reached the distribution system, the water system would alert the end-users through public notification systems and possibly issue do-not-use or boil-water orders. The utility would then have to decontaminate the system, which would require shutting down the distribution system while it is cleaned and disinfected. In that case, the utility would provide consumers with bottled water or distribute clean water from trucks called *water buffaloes*.

201. What about the danger of a chemical from a water utility being released into the air?

Just as with other security measures, utilities that use potentially dangerous chemicals, such as chlorine, have conducted risk assessments to determine what the effects of a chemical release would be and have taken precautions to safeguard their hazardous materials.

133

This includes protecting facilities, early detection of any dangerous chemical release, *scrubbing* equipment that would remove the chemical from the air before it can be released into the community and ensuring the safe delivery of hazardous materials.

Also, more and more utilities are switching to alternative disinfectants that are less hazardous.

202. Are water utilities prepared for natural disasters such as hurricanes and floods?

Water utilities have emergency response plans (ERPs) that provide them with a course of action in the event of a natural disaster or security breach. This plan will detail how potable water is to be delivered to consumers if something happens to the water supply or treatment and distribution facilities, and most utilities train on the ERP regularly.

That said, there is always the chance that something will happen that completely destroys a facility and prohibits utility personnel from working. If this were to happen, a utility could receive assistance from other water suppliers in neighboring communities, or even other states, through a program called *WARN*, which stands for Water and Wastewater Agency Response Networks. In the aftermath of Hurricane Katrina, for example, water department operators from Portland, Ore., assisted in the recovery of the New Orleans water and wastewater systems.

Fascinating Facts

Water is unique in that it is the only natural substance that is found in all three states—liquid, solid (ice), and gas (steam)—at the temperatures normally found on Earth. Earth's water is constantly interacting, changing, and in movement.
     — U.S. Geological Survey, *Water Science for Schools,* 2008

LIQUID

203. How much does water weigh?

One U.S. gallon (3.8 liters) of water weighs about 8.34 pounds (3.8 kilograms).

204. I have a typical residential lot of a quarter acre. If it rains 1 inch (25 millimeters), how much water falls on my lot?

About 7,000 U.S. gallons (26,500 liters), or nearly 30 tons (27 metric tons) of water.

205. Why does drinking water get bubbles in it after it is left out for a few hours?

Gases from the air, nitrogen and oxygen mostly, can dissolve in water (just like sugar dissolves in water). The amount of air in the water is different at different temperatures, however. Cold water can hold more dissolved air in it than warm water. A glass of water left at room temperature warms up after a while, and the dissolved air is forced out of the water into bubbles. This relieves the condition that is called *supersaturation*, which is what happens when the water contains more air than can be dissolved in the water.

206. Why does water swirl when it goes down the drain?

It's a common misconception that the earth spinning on its axis—the Coriolis effect—causes draining water to spin one direction in the Northern Hemisphere and another in the Southern Hemisphere. The rotation of the earth is a major factor in the motion of ocean currents and weather, but your bathtub or sink is too small to be substantially affected by this motion.

Instead, the swirling at the drain is caused by the same effect as the motions of a spinning ice skater. When the skater pulls her or his arms in close to the body, the skater spins faster. Water in a basin is never quite quiet, so when the water is pulled toward a drain, it's drawn down at an angle, and any residual rotation becomes larger on the broad surface and increasingly smaller toward the narrower drain opening, gaining momentum so you can see it, like the ice skater pulling her or his arms close to the body.

207. Why is water in the ocean blue?

Sunlight contains all colors of the rainbow, but when you see a color it is because that particular color of light bounces off the object and reaches your eye, and the other colors are absorbed by the pigment of the object. The sky is blue on a nice day because only blue light reflects off the particles in the atmosphere. The ocean is blue because as the blue light comes out of the sky, it reflects from the water and into your eye, fooling your eye into thinking the water is blue, when actually water has no color. The ocean absorbs red light, which leaves more blue light to reflect off the sea. Next time the sky is overcast, notice how the ocean loses most of its color because so little blue light is coming out of the sky.

Where the ocean is clear and shallow, the sea bottom can also influence the color, so the water will reflect different shades of blue, and often green. Swimming pools are usually painted blue to make the pool water appear blue.

208. Why is ocean water salty?

Ocean water contains a significant amount of dissolved sodium and chloride—salt—that comes from rivers and streams carrying tiny bits of rock that the flowing water has eroded on its way to the ocean. The erosion of the ocean floor also contributes to the salt content. Sea water is about 3.5 percent salt.

209. Why, then, aren't big lakes that are fed by rivers and streams salty?

All water has some salt content, although most taste buds are not astute enough to detect it. Fresh water lakes "turn over" at a relatively frequent rate, and rivers are fed by fresh snowmelt, rain and underground springs, which keep the water in them from accumulating salt.

210. Why should I wash my hands in hot water?

Many of the substances you are trying to get off your hands are somewhat greasy. Using hot water helps soften that greasy stuff on your hands, so soap or detergent can lift if off your skin. Think about it: when you put a container of liquid grease in the refrigerator, it hardens, but when it is warmed up, it melts. Hot water also helps the soap suds up, making it more effective at killing bacteria.

211. Is it true that a microwave oven heats food and beverages by flipping water molecules back and forth? How is that possible?

A water molecule has one oxygen atom with two hydrogen atoms connected to it, thus H_2O. The shape of the molecule is like a flat V, with the oxygen at the point and the two hydrogen atoms hanging off it. Oxygen has

two negative charges, and each hydrogen atom has a positive charge, so think of a water molecule as a little magnet, with one end negative and the other end positive. When the microwave is on, the waves affect the way the water molecules line up. Because the microwaves are produced by alternating current (AC), they change direction about 2 billion times each second. Each time the microwaves change direction, water molecules are flipped, too. Because water molecules are so small, trillions upon trillions of molecules are in any food or beverage you put in the microwave. All these molecules rapidly flipping back and forth create heat for defrosting, cooking and warming food and beverages.

212. I've heard of water exploding after it comes out of the microwave, causing serious burns. What can I do to avoid this?

Microwaved water and other liquids do not always bubble when they reach the boiling point. This happens when the water container has perfectly smooth surfaces and the water heats faster than the vapor bubbles can form. This superheated liquid will bubble up out of the cup when it is moved or when something is put into it, such as a spoon or tea bag. It is a rare occurrence, because most containers contain impurities that disrupt the perfect molecule alignment that is required.

To prevent this from happening, do not heat any liquid for more than two minutes per cup. After heating, let the cup stand in the microwave for thirty seconds before moving it or adding anything to it. Or put a wooden stir stick (no metal) in the water before you heat it.

213. Why does it take longer to cook things in boiling water when I am camping in the mountains?

The temperature at which water boils depends on the atmospheric pressure. Atmospheric pressure depends on the weight of the air above any location. At sea level, the weight of the atmosphere "pushes down" with a pressure of 14.7 pounds per square inch (101.3 kilopascals) and water boils at 212 °F (100 °C). Go up into the mountains, however, and the atmosphere is thinner, with less pressure. The lower the pressure, the lower the temperature at which water boils. The temperature at which water boils goes down about 2 °F (1 °C) for each 1,000 feet (305 meters) of elevation. If you are camping in the mountains at 10,000 feet (3,050 meters), water boils at only 201 °F (94 °C). At this lower temperature, food takes longer to cook.

214. Are raindrops really shaped like teardrops?

Small raindrops (with a diameter of less than 0.1 inch [0.25 cm]) are shaped like a sphere; larger ones are shaped more like a hamburger bun, according to the U.S. Geological Survey.[1] When raindrops become larger than a diameter of 0.4 inch (1 centimeter), they distort into a shape rather like a parachute with a tube of water around the base, and then they break up into smaller drops.

215. Why is it that water is referred to as "the universal solvent"?

Water is the universal solvent because it has the ability to break down almost any substance, be it a package of dried food that is mixed in it, or the Earth's surface as it flows over it. In our bodies, water can wash out the toxins captured by our kidneys. Scientifically speaking, a water molecule's simple, stable bond of two hydrogen atoms and one oxygen atom creates a neutral solution, neither acidic nor base, that breaks the bonds of larger, more complex molecules.

---

1   U.S. Geological Survey, http://ga.water.usgs.gov/edu/raindropshape.html

216.
I saw a TV demonstration in which a container was slowly filled with water, and the water rose to a level above the top of the container without spilling. How is this possible?

Water molecules at the surface form a thin skin on the top of the water, and this skin is what holds the water above the top of the container. This is called *surface tension*. If you break the skin, the water will spill down the side of the container.

Try this yourself. Fill a container partially with water. When the surface is still, carefully lay a sewing needle on the surface. The skin will hold the needle up and allow it to float on the surface.

217.
Why can't I pick up a piece of dirt from the floor with a dry finger, but if I wet my finger the dirt will stick to it?

This is a result of the same property of water as noted in the answer above. The surface tension of the liquid on your finger causes the dirt to cling to your finger.

SOLID

218. Why does ice float instead of sink?

Frozen water is less dense than liquid water because as water freezes, the hydrogen bonds connecting different water molecules line up to keep the negatively charged oxygen atoms apart, creating equal spaces between the molecules, freezing the water in a lattice-type formation. This results in a change in density that allows ice to float.

If this were not the case, ice would sink when it was formed and go to the bottom of all the lakes, ponds and reservoirs. Even icebergs would sink to the bottom of the ocean. When spring came, not enough heat would get to the bottom to melt the ice down there. So after a while, almost all the water on the earth would be frozen, and life as we know it would not have developed on the planet.

219. How do ice skates work to allow the skater to glide over the ice?

A thin film of liquid water covers the surface of any body of ice. How thick the liquid layer is depends on the surface temperature—the colder the ice, the thinner the layer. An ice skate blade bears the weight of the skater on a very small area, and when the skate blade passes over the ice, the increase in pressure and kinetic friction raise the temperature and melt the ice under the blade, creating an even thicker layer of water that has little resistance to the ice skate blade.

220. I saw a television science show experiment where one block of ice was hung from a wire looped around it, and another ice block was hung from a plastic line. Each had the same size weight hanging from it. The weight pulled the wire through the block of ice, but not the plastic line. What was going on there?

Wire is a better heat conductor than plastic, and it absorbs the heat from the ice that melts and refreezes under pressure from the weight. Plastic doesn't absorb and conduct heat nearly as well, so considerably less melting occurs.

221. What makes ice cubes cloudy?

Commercially made ice is stirred as it is being frozen; household ice is not. Without mixing, many more ice crystals form, and air is trapped in the ice. Light rays are distorted by these crystals and air, and this distortion gives home-frozen ice a cloudy appearance.

222. Why do ice cubes bulge from the top of the ice cube tray?

Unlike most things, which get smaller when they get colder, water gets bigger (expands) by 9 percent when it freezes. Because an ice cube tray has a bottom and four sides that don't move, ice bulges out of the open top when the water expands.

Also, because ice is expanded, it is less dense and lighter than water. So in the winter, ice floats on the surface of a water body while the water underneath stays liquid, providing organisms, including fish, with a place to survive during the cold weather.

223. Snow, sleet, freezing rain and hail— how are they different?

All precipitation begins as ice crystals in the upper atmosphere.

Snow falls through cold atmospheric layers that are consistently below freezing, giving the snowflake no opportunity to melt.

Sleet melts on the way down and refreezes into icy pellets before hitting the ground. Sleet usually occurs during the winter and is smaller than 0.30 inch (0.76 centimeter [cm]) in diameter.

Freezing rain, or drizzle, is snow that melts on the way down (like regular rain) and stays liquid, but freezes upon hitting the ground or other frozen surfaces. Freezing rain is associated with ice storms that coat trees, power lines and roads.

Hail is different. It is created during strong thunderstorms that contain strong updrafts. As the precipitation from a circular updraft first nears the ground, it is blown back up into the cold layer of sky where it picks up more moisture that freezes on the outside, making the hailstone a bit bigger. Then it starts down again. In very strong storms, the hail goes back up several times, getting bigger each time. When it finally gets so heavy that the updrafts cannot blow it upward anymore, it falls to the ground. If you pick up a large hailstone and cut through it, you will see rings as on a cut tree trunk, each ring indicating another trip up into the sky. Soft hail is called *graupel* or snow pellets. Hail usually is round, at least 0.20 inch (0.51 cm) in diameter, and occurs primarily in the spring and summer.

224. Is it true that all snowflakes are different, no two being exactly alike?

The eventual shape of the snowflake depends on the temperature of the original ice crystal, the temperature zones it falls through and how it falls, which is when it picks up additional crystals to become a flake. A perfectly symmetrical snowflake spins like a top on the way down.

Each snow crystal contains about 1 quintillion (1,000,000,000,000,000,000) molecules of water. Each molecule aligns itself differently, and all molecules are scattered randomly throughout the crystal, creating even more ways for the crystal to be different. Then, consider that the crystals fall at different rates and movements—spinning, arcing, flipping—depending on the wind currents and the various temperature changes they encounter as they fall. Some even pick up bits of dirt and debris in the air. All of this means it is unlikely that anyone would see two identical snowflakes.

225. Even when the temperature stays below freezing, the depth of the snow on the ground decreases. How is this possible?

When snow falls from the sky, it settles in layers that are full of air around the snowflakes. As the snowpack settles, it pushes some of the air out, and the snow on the ground becomes more compact. Also, wind and sun can affect the amount of snow on the surface, in a process similar to evaporation (liquid turning to gas), but known as *sublimation* (frozen matter turning to gas). Even if temperatures aren't warm enough to melt the snow, some will dissipate into water vapor, further decreasing the snowpack.

226. I saw a demonstration where someone put a black cloth on a snowbank on a cold, sunny day and the cloth sank into the snow. Why did that happen?

The sun sends heat and light to the earth as radiation. Many things reflect light into our eye, so we see them. Something black does not reflect any light, so we see it as black (the absence of reflected light). The same thing is true of heat. The black cloth does not reflect radiation and thus keeps both the light and the heat. This causes the black cloth to get warm, so it melts the ice beneath it. A white cloth won't melt the snow because it reflects both the light and heat, staying cool. Some people use a black hose to warm swimming pool water in sunny areas. The black hose lies out in the sun, collects the heat from sunlight and warms the water being pumped slowly through it.

227. Why is the ice in glaciers blue?

The weight of the glacier produced tremendous pressure on the ice for a long, long time, which changed the nature of the tiny ice crystals so that they only reflect short, blue light wavelengths, while absorbing long, red wavelengths. The longer the path that light must travel through the ice, the bluer the ice appears. This does not happen when you make ice in the refrigerator, because this ice is not under high pressure.

228. We often use artificial cold packs in our coolers to keep things cold. How do they work?

Commercial products are about 98 percent water, but they have added chemicals—usually ammonium chloride—to lower the freezing point. Because the chemical reaction between the water and the chemical lowers the temperature of the pack to below 32 °F (0 °C), the packs can absorb heat from other objects, and thus keep food and drinks cold. This idea is not new. Before freezers were common, ice cream was made by surrounding cream-filled containers with water, ice and salt. The added salt lowered the temperature of the surrounding mixture and drew heat out of the cream until it froze. When salt or calcium is used to de-ice pavements or sidewalks, the same principle applies.

Reusable gel packs contain water mixed with a refrigerant gel that prevents the water from completely freezing, so the pack can conform to the shape of the object it is next to.

GAS

229. What is dry ice?

Dry ice is a solid form of carbon dioxide (CO_2). Unlike water, CO_2 does not have a liquid form and can only be a gas or a solid. Water freezes at 32° F (0 °C), but because it is originally a gas, CO_2 must be much colder to form a solid. For CO_2 to be come dry ice, it must be cooled down to −109.3 °F (−78.5 °C).

It is called dry ice because no liquid is involved. When dry ice warms up, it doesn't melt into a liquid, but escapes into the air as a gas by the process of sublimation. Dry ice is often used to make smoke or fog in stage productions by placing it in a bucket of water, which creates low-sinking dense clouds of fog.

230. If we can't see gas, why can we see steam from a boiling teakettle or a hot cup of soup?

It's true that pure vaporized water is a completely invisible gas. But the steam you see above a hot, wet surface is really a mist or vapor that occurs when the gas mixes with the cooler air, which slows the gas molecules down and combines them with other water molecules, condensing them and forming tiny visible droplets of liquid water.

231. Why can we see our breath when it is cold outside?

The effect is almost the same as a boiling teakettle; the air from your lungs is full of moisture, and when your warm breath hits the freezing air, it condenses into a visible liquid, or if it's very cold, into frozen crystals.

Appendix A

Acronym List

ANSI	Americn National Standards Institute
AWWA	American Water Works Association
BDL	below detection limits
CCL	Contaminant Candidate List
CCR	Consumer Confidence Report
CDC	Centers for Disease Control and Prevention
CDW	Federal–Provincial–Territorial Committee on Drinking Water (Canada)
cm	centimeter
CO_2	carbon dioxide
CWA	Clean Water Act
DBP	disinfection by-product
EDC	endocrine disrupting compounds
ERP	emergency response plan
FDA	U.S. Food and Drug Administration
HAA5	five haloacetic acids
HDPE	high-density polyethylene
IRIS	Integrated Risk Information System
ISAC	Information Sharing and Analysis Center
GAC	granular activated carbon

GWR   Ground Water Rule

kPa   kilopascals

$KMnO_4$ Potassium permanganate

MAC   maximum acceptable concentration

MCL   maximum contaminant level

MCLG   maximum contaminant level goal

mg/L   milligrams per liter

ND   not detected

NRC   U.S. Nuclear Regulatory Commission

NSF   National Science Foundation

POE   point-of-entry

POU   point-of-use

PPCPs   pharmaceuticals and personal care products

ppm   parts per million

psi   pounds per square inch

PVC   polyvinyl chloride

RO   reverse osmosis

SMCL   secondary maximum contaminant levels

SDWA   Safe Drinking Water Act

THMs   trihalomethanes

UCMR   Unregulated Contaminant Monitoring Rule

UV   ultraviolet

USEPA   U.S. Environmental Protection Agency

USP   United States Pharmacopeia

WARN   Water & Wastewater Agency Response Networks

WHO   World Health Organization

Appendix B

Complete
List of Questions

HEALTH

1. Is my water safe to drink?
2. Is potable water the same as safe water?
3. How much water should I drink to stay healthy?
4. Must all my water intake be plain water or do drinks made with water count?
5. Does drinking water contain calories, fat, sugar, caffeine or cholesterol?
6. Can drinking water help me lose weight?
7. Does drinking ice water help burn fat?
8. Is there a connection between drinking water and healthy skin?
9. How long can a human go without water?
10. Is it true that you can drink too much water?
11. Does anyone actually get sick from drinking tap water?
12. Is tap water safer in one area of a community as compared with another?
13. People are allowed to swim and go boating in our water supply reservoir. Should I worry about this?
14. Should I use tap water for my baby's formula?
15. Is tap water suitable for use in a home kidney dialysis machine?
16. Is it safe to take a drink from my garden hose?
17. Is it safe to drink water from a drinking fountain?

18. Are in-store water dispensers safe?
19. Should I buy drinking water from a vending machine?
20. Is there salt in my water?

MICROBIAL CONTAMINANTS
21. What types of living organisms do I need to be concerned about in water?
22. How are germs kept out of my drinking water?
23. If I am concerned about my water, how can I kill the germs in it?
24. Can flu viruses be spread by water?
25. What are *Cryptosporidium* and cryptosporidiosis?
26. Are all water systems at risk from *Cryptosporidium*?
27. Is drinking water the only source of *Cryptosporidium*?
28. My water supplier found some of the microbes that cause cryptosporidiosis in the source water. Will I get sick if I drink water from the tap?
29. Could my drinking water transmit the AIDS virus?

CHEMICAL AND MINERAL CONTAMINANTS
30. Are chemicals found naturally in drinking water nontoxic?
31. What are organic chemicals? Are they dangerous?
32. Where can I find information about how environmental exposures affect human health?
33. The movie "Erin Brockovich" focuses on the chemical chromium-6. What is that and is it a threat to other water supplies?
34. Do hazardous wastes contaminate drinking water?

Pharmaceuticals
35. Are there drugs in my water? How do they get there?
36. Should I be concerned about pharmaceuticals and personal care products in the water?
37. What are endocrine disrupting compounds (EDCs)?
38. If flushing drugs down the toilet causes environmental problems, how should I dispose of unwanted medication?

Arsenic

39. What is arsenic and how does it get in my drinking water?
40. What are the health effects of arsenic exposure?
41. Is arsenic regulated in drinking water?
42. What home treatment systems best remove arsenic?

Atrazine

43. What is atrazine?
44. Can atrazine get into the water and affect human health?
45. Is atrazine regulated?

Lead and Copper

46. How does lead get into drinking water?
47. How can I get lead or copper out of my drinking water?
48. What are the health effects of drinking water with lead in it?

Nitrates

49. How do nitrates and pesticides get into my drinking water? What health problems do they cause?

Perchlorate

50. What is perchlorate?
51. Is perchlorate regulated?
52. How can perchlorate be removed if it gets in my drinking water?

Radon and Radium

53. What is radon and is it harmful in drinking water?
54. My private well has high levels of radon. How can I make sure my drinking water is safe?
55. What is radium and how is it treated?

WATER TREATMENT

56.   Are all chemicals in my drinking water bad for me?

Chlorine

57.   Is water with chlorine in it safe to drink?
58.   Is there a link between chlorine and cancer?
59.   Should I be concerned about the chlorine in the water I use for bathing or showering?
60.   What else is used to kill the germs in water?

Aluminum

61.   I hear aluminum is used to treat drinking water. Is this a problem? Does it cause Alzheimer's disease?

Fluoride

62.   Is the fluoride in my drinking water safe?
63.   Will I lose the benefits of fluoride in my drinking water if I install a home treatment device or drink bottled water?

TRAVEL

64.   If I travel overseas, is the tap water safe to drink?
65.   Is there something I can take with me when I travel to purify water?
66.   When I travel to a different place in this country, sometimes I have an upset stomach for a couple of days. Is this because something is wrong with the water?
67.   I've heard that some countries use solar rays to disinfect their drinking water. Is that true?
68.   How is water quality protected on airplanes?
69.   We sometimes hear about waterborne illness outbreaks on cruise ships. How is water quality protected on cruise ships?
70.   How is drinking water on cruise ships treated?
71.   Do I need more water if I'm moving or traveling to a higher altitude?
72.   Is it okay for campers, hikers and backpackers to drink water from remote streams?

73. What can campers, hikers and backpackers do to treat stream water to make it safe to drink?

TASTE, ODOR AND APPEARANCE

74. Can I tell if my drinking water is okay by just looking at it, tasting it, or smelling it?
75. Why does my drinking water taste or smell "funny"? Will this smelly water make me sick?

Flat-tasting Water

76. What can I do if my drinking water tastes "funny"?
77. Why do the ice cubes from my freezer used to cool my water make the water taste funny?
78. White stuff appears in the glass as ice cubes from my freezer melt. What is the white stuff and where does it come from?
79. Why does drinking water often look cloudy when first drawn from a faucet?
80. My drinking water is reddish or brown. What causes this?
81. My drinking water is dark in color, nearly black. What causes this?

IN THE HOME

82. How much water does one person use each day?
83. How long can I store drinking water?
84. How much water should I store for emergencies?
85. How do I treat my water in an emergency?
86. Should I use hot water from the tap for cooking?
87. Should I use hot water from the tap to make baby formula?
88. Is it okay to heat water for coffee or tea in a micro-wave in a foam cup?
89. Is water that comes out of a dehumidifier safe to drink?

90. How is the water from my refrigerator door different from my tap water?

HOME PLUMBING
91. What should I do to avoid cold-weather problems with my pipes?
92. Where do I find my home's master valve?
93. How can I tell if I have leaks in my home plumbing system?
94. How can I tell if my toilet is leaking?
95. Why are there aerators on home water faucets?
96. Why do hot water heaters fail?
97. What causes the banging or popping noise that some water heaters, radiators and pipes make?
98. There is a blue-green stain where my water drips into my sink. What causes this?
99. Why does my water sometimes have sand in it?

HOME TREATMENT
100. Should I install home water treatment devices?
101. How do I know which type of POU/POE to use?
102. I heard about a water treatment device that uses an electromagnet to treat water. Does this work?
103. Is distilled water the "perfect" drinking water?

HARD WATER AND SOFTENING
104. What is "hard" water?
105. Should I install a water softener in my home?
106. Does softened drinking water have any negative health effects?
107. I have a water softener, but I still get spots on my bathroom tile and dishware. Why?
108. How can I get rid of the precipitate deposits on my coffee or tea pot and showerhead?
109. Some of my clear glassware comes out of the dishwasher with a rainbow sheen on it. What causes this?

BOTTLED WATER

110. Should I buy bottled water?
111. Is bottled water regulated?
112. What do the labels on bottled water mean?
113. Is bottled water okay to store?

DOWN THE DRAIN

114. Where does the water go when it goes down the drain?
115. What can I safely pour down the sink or into the toilet?
116. How do I safely dispose of pharmaceuticals?
117. I have a septic tank. Should I take any special precautions?

FISH AND PLANT LIFE

118. How should I fill my fish aquarium?
119. I have trouble keeping fish alive in my fish pond. Is there anything I can do?
120. When I try to root a plant or grow flowers from a bulb in my house, the water looks terrible after a while. What will prevent that?
121. Roses, azaleas, camellias and rhododendron all require acid conditions. How should I adjust the acid content of my plant water?
122. I live in a very hard water area and I have a water softener. My plants don't seem to like my tap water. What can I do?
123. What causes the whitish layer on the soil of my potted plants?

COSTS

124. What is the cost of the water I use in my home?
125. How does the water utility know how much water I use in my home?
126. How does the water company know that my water meter is correct?
127. We had a conservation drive in our area and everyone cooperated. Then our water rates went up. Why?

SOURCES

QUALITY

145. I live downstream from a nuclear power plant. Should I worry about radioactivity in my drinking water?
146. How can I help prevent pollution of drinking water sources?
147. I have a private well. Where can I get my water tested, and what should it be tested for?
148. How can I protect my private water supply?

DISTRIBUTION

149. Are the pipes that carry drinking water from the treatment plant to my home clean?
150. I have seen work crews cleaning water mains, and the water they flush out looks terrible. How can the water be safe if the pipes are so dirty?
151. What are distribution pipes made from?
152. Why are fire hydrants sometimes called fire plugs?
153. We don't use much water for drinking. Why does all the rest need to be treated so extensively? This seems unnecessarily expensive.
154. How does a water utility detect a major leak in the distribution piping system?
155. If leaks are such a waste of water, why can't the water utility just fix all of them?
156. Fixing a broken water pipe looks like a dirty job. How is the inside of the pipe cleaned afterward?
157. What causes low water pressure?
158. What are cross-connections?
159. Why is some drinking water stored in large tanks high above the ground?

CONSERVATION

160. What indoor home activity uses the most water?
161. How much water can we save by installing new fixtures?
162. Why can't I just put a brick in my toilet instead of replacing it?

163. Before I replace my showerhead, how do I measure how fast my shower is using water?
164. What about outside use—how can I conserve water there?
165. How should I irrigate my lawn and garden to avoid wasting water?
166. Can I save water by changing my landscaping to Xeriscape?
167. I heard that some water conservation measures are actually required by law. Is this true?
168. Which uses more water, a tub bath or a shower?
169. I leave the water running while I brush my teeth. Does this waste much water?
170. I use a lot of water in the kitchen. How can I conserve there?
171. Many water quality problems in the home—lead, red water, sand in the system and so forth—are cured by flushing the system. Isn't that a waste of water?
172. My water faucet drips. Should I bother to fix it?
173. How concerned should I be about a leaky toilet?
174. I have a private well. Why should I conserve water?
175. Why do we still have a water shortage when it's been raining at my house?
176. During water shortages, shouldn't decorative fountains be turned off?
177. Do restaurants actually conserve water by not serving it to customers?
178. Since the amount of water on the globe isn't changing, and the water in my area is plentiful, why should I conserve?
179. What is gray water? Can we use it?
180. Our water utility has opened a reclaimed water plant in our town. What is reclaimed water? Can we drink it?
181. Is reclaimed water regulated?

REGULATIONS, REPORTING AND WATER SECURITY

182. Is the quality of drinking water regulated by legislation?
183. What about state or provincial regulations?
184. Is water that meets government drinking water standards absolutely safe?
185. How do regulatory agencies choose the standard for a chemical in drinking water?
186. How do we know what the allowable level of a contaminant is?
187. Most federal standards are written like this: "Selenium—0.05 mg/L." What does "mg/L" mean?
188. What is the Clean Water Act?
189. How do regulatory agencies choose the contaminants to regulate?
190. If most tap water is safe, why are engineers and scientists still doing so much research and why is the government thinking about more regulations?
191. Must all surface water supplies be filtered?
192. What is the Ground Water Rule?
193. How is my water tested and who tests it?
194. Can water systems be excused from monitoring for some contaminants?
195. How do I find out if my water is safe to drink?
196. What information should I look for in the water quality report from my supplier?
197. I've received a notice from my water utility telling me that something is wrong. What's that all about? What is a boil-water order?
198. I live in an apartment and don't get a water bill. How will I know if there are any problems with my tap water?
199. What are the chances of a water system being the object of a terrorist attack?
200. What would happen if somebody intentionally put a dangerous chemical in our water?

201. What about the danger of a chemical from a water utility being released into the air?
202. Are water utilities prepared for natural disasters such as hurricanes and floods?

FASCINATING FACTS

LIQUID

203. How much does water weigh?
204. I have a typical residential lot of a quarter acre. If it rains 1 inch (25 millimeters), how much water falls on my lot?
205. Why does drinking water get bubbles in it after leaving it out for a few hours?
206. Why does water swirl when it goes down the drain?
207. Why is water in the ocean blue?
208. Why is ocean water salty?
209. Why, then, aren't big lakes that are fed by rivers and streams salty?
210. Why should I wash my hands in hot water?
211. Is it true that a microwave oven heats food and beverages by flipping water molecules back and forth? How is that possible?
212. I've heard of water exploding after it comes out of the microwave, causing serious burns. What can I do to avoid this?
213. Why does it take longer to cook things in boiling water when I am camping in the mountains?
214. Are raindrops really shaped like teardrops?
215. Why is it that water is referred to as "the universal solvent"?
216. I saw a TV demonstration in which a container was slowly filled with water, and the water rose to a level above the top of the container without spilling. How is this possible?
217. Why can't I pick up a piece of dirt from the floor with a dry finger, but if I wet my finger the dirt will stick to it?

SOLID

218. Why does ice float instead of sink?
219. How do ice skates work to allow the skater to glide over the ice?
220. I saw a television science show experiment where one block of ice was hung from a wire looped around it, and another ice block was hung from a plastic line. Each had the same size weight hanging from it. The weight pulled the wire through the block of ice not the plastic line. What was going on here?
221. What makes ice cubes cloudy?
222. Why do ice cubes bulge from the top of the ice cube tray?
223. Snow, sleet, freezing rain and hail—how are they different?
224. Is it true that all snowflakes are different, no two being exactly alike?
225. Even when the temperature stays below freezing, the depth of the snow on the ground decreases. How is this possible?
226. I saw a demonstration where someone put a black cloth on a snowbank on a cold, sunny day and the cloth sank into the snow. Why did that happen?
227. Why is the ice in glaciers blue?
228. We often use artificial cold packs in our coolers to keep things cold. How do they work?

GAS

229. What is dry ice?
230. If we can't see gas, why can we see steam from a boiling teakettle or a hot cup of soup?
231. Why can we see our breath when it is cold outside?

Appendix C

Natural Inorganic Chemicals Found in Drinking Water

This material is adapted from: Sandeen, W. G., "Groundwater Resources of Rusk County, Texas," U.S. Geological Survey, Open File Report 83-757 (1983). Note: Secondary regulations of the U.S. Environmental Protection Agency mentioned below are nonenforced recommendations.

ALKALINITY

Description and Sources
Alkalinity is a measure of the capacity of a water to neutralize a strong acid, usually to pH of 4.2. Alkalinity in natural waters usually is caused by the presence of bicarbonate and carbonate ions and to a lesser extent by hydroxide and minor acid radicals such as borates, phosphates and silicates. Carbonates and bicarbonates are common to most natural waters because of the abundance of carbon dioxide and carbonate minerals in nature. The alkalinity of natural waters varies widely but rarely exceeds 400 to 500 mg/L as $CaCO_3$.

Effect on Drinking Water
Alkaline waters may have a distinctive unpleasant taste.

CALCIUM (Ca)

Description and Sources

Calcium is widely distributed in the common minerals of rocks and soils and is the principal cation in many natural fresh waters, especially those that contact deposits or soils originating from limestone, dolomite, gypsum and gypsiferous shale. Calcium concentrations in freshwaters usually range from zero to several hundred mg/L. Larger concentrations are not uncommon in waters in arid regions, especially in areas where some of the more soluble rock types are present.

Effect on Drinking Water

Calcium contributes to the total hardness of water. Small concentrations of calcium carbonate combat corrosion of metallic pipes by forming protective coatings. Calcium in domestic water supplies is objectionable because it tends to cause incrustations on cooking utensils and water heaters and increases soap or detergent consumption in waters used for washing, bathing and laundering.

CHLORIDE (Cl⁻)

Description and Sources

Chloride is relatively scarce in the earth's crust but is the predominant anion in sea water, most petroleum-associated brines, and in many natural freshwaters, particularly those associated with marine shales and evaporites. Chloride salts are very soluble and once in solution tend to stay in solution. Chloride concentrations in natural waters vary from less than 1 mg/L in stream runoff from humid areas to more than 100,000 mg/L in groundwaters and surface waters in arid areas. The discharge of human, animal or industrial wastes and irrigation return flows may add significant quantities of chloride to surface and groundwaters.

Effect on Drinking Water

Chloride may impart a salty taste to drinking water and may accelerate the corrosion of metals used in water-supply systems. According to the National Secondary Drinking Water Regulations of the U.S. Environmental Protection Agency, the maximum contaminant level of chloride for public water systems is 250 mg/L.

DISSOLVED SOLIDS

Description and Sources

Theoretically, dissolved solids are dry residues of the dissolved substances in water. In reality, the term "dissolved solids" is defined by the method used in the determination. In most waters, the dissolved solids consist predominantly of silica, calcium, magnesium, sodium, potassium, carbonate, bicarbonate, chloride and sulfate, with minor or trace amounts of other inorganic and organic constituents. In regions of high rainfall and relatively insoluble rocks, waters may contain dissolved-solids concentrations of less than 25 mg/L; but saturated sodium chloride brines in other areas may contain more than 300,000 mg/L.

Effect on Drinking Water

Dissolved-solids values are used widely in evaluating water quality and in comparing waters. The following classifications based on the concentrations of dissolved-solids commonly is used by the U.S. Geological Survey.

Classification	Dissolved-solids concentration (mg/L)
Fresh	< 1,000
Slightly saline	1,000–3,000
Moderately saline	3,000–10,000
Very saline	10,000–35,000
Brine	> 35,000

The National Secondary Drinking Regulations (U.S. Environmental Protection Agency) set a dissolved-solids concentration of 500 mg/L as the maximum contaminant level for public water systems. This level was set primarily on the basis of taste thresholds and potential physiological effects, particularly the laxative effect on unacclimated users. Although drinking waters containing more than 500 mg/L are undesirable, such waters are used in many areas where less mineralized supplies are not available without any obvious ill effects.

FLUORIDE (F⁻)

Description and Sources

Fluoride is a minor constituent of Earth's crust. The calcium fluoride mineral fluorite is a widespread constituent of resistate sediments and igneous rocks, but its solubility in water is negligible. Fluoride commonly is associated with volcanic gases, and volcanic emanations may be important sources of fluoride in some areas. The fluoride concentration in fresh surface waters usually is less than 1 mg/L, but larger concentrations are not uncommon in saline water from oil wells, groundwater from a wide variety of geologic terrain, and water from areas affected by volcanism.

Effect on Drinking Water

Fluoride in drinking water decreases the incidence of tooth decay when the water is consumed during the period of enamel calcification. Excessive quantities in drinking water consumed by children during the period of enamel calcification may cause a characteristic discoloration (mottling) of the teeth. Thus, the EPA has an upper allowable limit for fluoride in drinking water.

HARDNESS

Description and Sources
Hardness of water is attributable to all polyvalent metals but principally to calcium and magnesium ions. Water hardness results naturally from the solution of calcium and magnesium, both of which are widely distributed in common minerals of rocks and soils. Hardness of waters in contact with limestone commonly exceeds 200 mg/L. In waters from gypsiferous formations, a hardness of 1,000 mg/L is not uncommon.

Effect on Drinking Water
Excessive hardness of water for domestic use is objectionable because it causes incrustations on cooking utensils and water heaters and increased soap or detergent consumption.

IRON (Fe)

Description and Sources
Iron is an abundant and widespread constituent of many rocks and soils. Iron concentrations in natural waters are dependent upon several chemical processes including oxidation and reduction; precipitation and solution of hydroxides, carbonates and sulfides; complex formation, especially with organic material; and the metabolism of plants and animals. Dissolved-iron concentrations in oxygenated surface waters seldom are as much as 1 mg/L. Some groundwaters, unoxygenated surface waters such as deep waters of stratified lakes and reservoirs, and acidic waters resulting from discharge of industrial wastes or drainage from mines may contain considerably more iron.

Corrosion of iron casings, pumps, and pipes may add iron to water pumped from wells.

Effect on Drinking Water

Iron is an objectionable constituent in water supplies for domestic use because it may adversely affect the taste of water and beverages and stain laundered clothes and plumbing fixtures. According to the USEPA National Secondary Drinking Water Regulations, the maximum contamination level of iron for public water systems is 0.3 mg/L.

MAGNESIUM (Mg)

Description and Sources

Magnesium ranks eighth among the elements in order of abundance in Earth's crust and is a common constituent in natural water.

Ferromagnesian minerals in igneous rock and magnesium carbonate in carbonate rocks are two of the more important sources of magnesium in natural waters. Magnesium concentrations in freshwaters usually range from zero to several hundred mg/L; but larger concentrations are not uncommon in waters associated with limestone or dolomite.

Effect on Drinking Water

Magnesium contributes to the total hardness of water. Large concentrations of magnesium are objectionable in domestic water supplies because they can exert a cathartic and diuretic action upon unacclimated users and increase soap or detergent consumption in waters used for washing, bathing and laundering.

MANGANESE (Mn)

Description and Sources

In chemical behavior and occurrence in natural water, manganese resembles iron. Manganese is much less abundant in rocks, however. As a result, the concentration of manganese in water is generally less than that of iron. Under reducing conditions (the lack of oxygen dissolved in water) in water containing dissolved carbon dioxide, manganese dissolves as manganous ion. This sometimes occurs in groundwaters and in water near the bottom of lakes and reservoirs.

Manganous ion is more stable in water in the presence of dissolved oxygen than ferrous (reduced) iron under similar circumstances.

Manganese concentrations greater than 1 mg/L may result where manganese-bearing minerals are attacked by water under reducing conditions or where some types of bacteria are active.

Effect on Drinking Water

Manganese is an essential trace element for humans. It plays an important role in many enzyme systems. Chronic toxicity has not been reported. With surface waters averaging less that 0.05 mg/L of manganese in several surveys, the potential harm from this source is virtually nonexistent. The main problem with manganese in drinking water has to do with undesirable taste and discoloration (black) of the water. The EPA secondary drinking water standard for manganese in drinking water is 0.05 mg/L to prevent these problems.

NITROGEN (N)

Description and Sources

A considerable part of the total nitrogen of Earth is present as nitrogen gas in the atmosphere. Small amounts of nitrogen are present in rocks, but the element is concentrated to a greater extent in soils or biological material. Nitrogen is a cyclic element and may occur in water in several forms. The forms of greatest interest in water, in order of increasing oxidation state, include organic nitrogen, ammonia nitrogen (NH_3-N), nitrite nitrogen (NO_2-N), and nitrate nitrogen

(NO$_3$-N). These forms of nitrogen in water may be derived naturally from the leaching of rocks, soils and decaying vegetation; from rainfall; or from biochemical conversion of one form to another. Other important sources of nitrogen in water include effluent from wastewater treatment plants, septic tanks and cesspools and drainage from barnyards, feedlots and fertilized fields. Nitrate is the most stable form of nitrogen in an oxidizing environment and is usually the dominant form of nitrogen in natural waters. Significant quantities of reduced nitrogen often are present in some groundwaters, deep unoxygenated waters of stratified lakes and reservoirs.

Effect on Drinking Water

Nitrate and nitrite are objectionable in drinking water because of the potential risk to bottle-fed infants for methemoglobinemia, a sometimes fatal illness related to the impairment of the oxygen-carrying ability of the blood.

pH

Description and Sources

The pH of a solution is a measure of its hydrogen ion activity. By definition, the pH of pure water at a temperature of 25 °C is 7.00. Natural waters contain dissolved gases and minerals, and the pH may deviate significantly from that of pure water. Rainwater not affected significantly by atmospheric pollution generally has a pH of 5.6 because of the solution of carbon dioxide from the atmosphere. The pH range of most natural surface waters and groundwaters is about 6.0 to 8.5. Many natural waters are slightly basic (pH >7.0) because of the prevalence of carbonates and bicarbonates, which tend to increase the pH.

Effect on Drinking Water

The pH of a domestic water supply is significant because it may affect taste, corrosion potential and water-treatment processes. Acidic waters may have a sour taste and cause corrosion of metals and concrete. The USEPA National Secondary Drinking Water Regulations set a pH range of 6.5 to 8.5 as the maximum contaminant level for public water systems.

POTASSIUM (K)

Description and Sources

Although potassium is only slightly less common than sodium in igneous rocks and is more abundant in sedimentary rocks, the concentration of potassium in most natural waters is much smaller than the concentration of sodium. Potassium is liberated from silicate minerals with greater difficulty than sodium and is more easily adsorbed by clay minerals and reincorporated into solid weathering products. Concentrations of potassium of more than 20 mg/L are unusual in natural fresh waters, but much larger concentrations are not uncommon in brines or in water from hot springs.

Effect on Drinking Water

Large concentrations of potassium in drinking water may act as a cathartic, but the range of potassium concentrations in most domestic supplies seldom causes these problems.

SILICA (SiO$_2$)

Description and Sources

Silica ranks second only to oxygen in abundance in Earth's crust. Contact of natural waters with silica-bearing rocks and soils usually results in a concentration range of about 1 to 30 mg/L; but concentrations as large as 100 mg/L are common in waters in some areas.

Effect on Drinking Water

Silica in some domestic water supplies may inhibit corrosion of iron pipes by forming protective coatings.

SODIUM (Na)

Description and Sources

Sodium is an abundant and widespread constituent
of many soils and rocks and is the principal cation in
many natural waters associated with argillaceous sediments, marine shales and evaporites, and in seawater.
Sodium salts are very soluble and once in solution tend
to stay in solution. Sodium concentrations in natural
waters vary from less than 1 mg/L in stream runoff
from areas of high rainfall to more than 100,000 mg/L
in groundwaters and surface waters associated with
halite deposits in arid areas. In addition to natural
sources of sodium, wastewater, industrial effluents,
oilfield brines and de-icing salts may contribute sodium
to surface and groundwaters.

Effect on Drinking Water

Sodium in drinking water may be harmful to persons
suffering from cardiac, renal and circulatory diseases
and to women with toxemias of pregnancy. Large
sodium concentrations are toxic to most plants.

SPECIFIC CONDUCTANCE

Description and Sources

Specific conductance is a measure of the ability of water
to transmit an electrical current and depends on the
concentrations of ionized constituents dissolved in the
water. Many natural waters in contact only with gran-
ite, well-leached soil, or other sparingly soluble material
have a low conductance.

Effect on Drinking Water

The specific conductance is an indication of the degree of mineralization of a water and may be used to estimate the concentration of dissolved solids in the water.

SULFATE (SO$_4$)

Description and Sources

Sulfur is a minor constituent of Earth's crust but is widely distributed as metallic sulfides in igneous and sedimentary rocks. Weathering of metallic sulfides such as pyrite by oxygenated water yields sulfate ions to the water. Sulfate is also dissolved from soils and evaporite sediments containing gypsum or anhydrite. The sulfate concentration in natural fresh waters may range from zero to several thousand mg/L. Drainage from mines may add sulfate to waters by virtue of pyrite oxidation.

Effect on Drinking Water

Sulfate in drinking water may impart a bitter taste and act as a laxative on unacclimated users. According to the EPA National Secondary Drinking Water Regulations the maximum contaminant level of sulfate for public water systems is 250 mg/L.

ZINC (Zn)

Description and Sources

Generally, in streams and rivers, zinc is concentrated in sediments, but concentrations are quite low in running filtered water. In areas of soft, acidic water, however, pickup in the distribution system has been noted when comparing water samples from the treatment plant with samples at consumer's taps. This may result from the water flowing through galvanized iron pipes. In rocks, zinc is most commonly present in the form of the sulfide sphalerite, which is the most important zinc ore. Zinc may replace iron or magnesium in certain minerals and it may be present in carbonate sediments. In the weathering process, soluble compounds of zinc are formed and the presence of at least traces of zinc in water is common.

Effect on Drinking Water

Zinc is considered an essential trace element in human and animal nutrition. The recommended daily dietary allowances for zinc are as follows: adults, 15 milligrams each day, growing children over a year old, 10 milligrams each day, and additional supplements during pregnancy and breast feeding (check with your doctor). As far as drinking water is concerned, the EPA secondary drinking water standard of 5 mg/L, assuming the common intake of two liters of water each day, would result in an intake of 10 milligrams of zinc each day, which is less than the estimated adult dietary requirement for zinc.

Concentrations of 40 milligrams in a liter of water gives the water a strong metallic taste (the technical description is astringent).

Appendix D

Pronunciation Guide

alum — AL-um

astringent — a-STRIN-jent

bacteriophage — back-TEER-ee-o-fahj
(rhymes with garage)

carcinogen — car-SIN-o-jin

catalytic — cat-a-LIT-ic

chloramine — KLOR-uh-mean

coriolis — kor-e-O-lis

corrosivity — kor-roh-SIV-i-tee

cryptosporidiosis — KRIP-toe-spo-rid-ee-OH-sus

Cryptosporidium — KRIP-toe-spor-ID-ee-um

cyanobacteria — SIGH-an-o-back-TEER-ee-ah

disinfectant — dis-in-FEC-tant

diuretic — die-u-RET-ic

endocrine — EN-doe-krin

enterococci — EN-tuh-ro-COCK-sigh

Escherichia — es-chuh-RIK-ee-ah

Giardia — jee-R-dee-ah

giardiasis — jee-R-dee-ah-sis

haloacetic — hay-loh-ah-SEED-ic

hyponatremia — HI-po-ni-TREE-me-ah

methemoglobinemia — met-hee-muh-glo-buh-NEE-me-ah

methyl-tertiary-butyl-ether —

      METH-uhl TUR-she- air-ee-BYU-tuhl-thur

microcystin — mi-kro-SIS-tin

microorganism — mi-kro-OR-gan-iz-m

oocyst — OH-oh-sist

osmosis — ahs-MO-sis

pathogen — PATH-o-jin

plutonium — plu-TONE-ee-um

potable — POTE-uh-bul (rhymes with floatable)

potassium — po-TAS-ee-um

precipitate — pree-SIP-uh-tate

sublimation — sub-la-MAY-shon

trihalomethane — tri-HA-LO-meth-ane

turbidity — tur-BID-it-ee

Xeriscape — ZER-uh-scape

About the Authors

Dr. James M. Symons, Retired, is the Cullen Distinguished Professor Emeritus of Civil Engineering University of Houston, and holds a doctor of science degree from Massachusetts Institute of Technology. He is a diplomate of the American Academy of Environmental Engineers and a member of the National Academy of Engineering and American Water Works Association (AWWA). In addition to his academic career, he has worked for the U.S. Environmental Protection Agency and U.S. Public Health Service and has received numerous research and publications awards and honors. He is also the editor emeritus of *The Water Dictionary, A Comprehensive Reference of Water Terminology*, published by AWWA.

Gay Porter De Nileon, MPA, is the publications manager for American Water Works Association, where she has worked in various editorial positions since 1993. She holds a master's degree in public administration from the University of Colorado – Denver and a bachelor of science in journalism from University of Colorado – Boulder. She has received numerous writing and editing awards, including one for her work on *Water Adventures Around the World, An Activity Book*, also published by AWWA.

Allen Panek is a water and wastewater operations consultant and a member of the AWWA Public Affairs and Administrative and Policy Councils. He was

Assistant Director of Public Utilities–Water & Wastewater for the City of Naperville, Illinois, and a staff member at Argonne National Laboratory, where he authored numerous technical publications in the field of environmental science/engineering. Allen has a bachelor's degree in chemistry from North Central College and has completed graduate-level coursework at Roosevelt University, Chicago. He holds the highest level of Illinois operator certification in both wastewater treatment and public water supply.

Jackie Glover serves as the Public Affairs and Diversity Manager for Park Water Company in Downey, Calif. She has been actively involved with AWWA since 1992, starting out as a member of the Customer Service Activities Committee. She is the past chair of the California/ Nevada Section of AWWA and of the Communications and Customer Relations Committee. She is also an AWWA QualServe peer reviewer. Jackie has a bachelor's degree in economics from Southern Methodist University in Dallas.

Index

CPSIA information can be obtained at www.ICGtesting.com
Printed in the USA
LVOW04s2025171114

414122LV00020B/1123/P